EPIGENETIC LANDSCAPES

EPIGENETIC

LANDSCAPES

Drawing
as
Metaphor

SUSAN MERRILL SQUIER

Duke University Press — Durham and London — 2017

Cover designed by Matthew Tauch;
Interior designed by Heather Hensley
Typeset in Chaparral Pro by Westchester Publishing Services

Library of Congress Cataloging-in-Publication Data
Names: Squier, Susan Merrill, author.
Title: Epigenetic landscapes : drawings as metaphor /
Susan Merrill Squier.
Description: Durham : Duke University Press, 2017. |
Includes bibliographical references and index.
Identifiers: LCCN 2017019584 (print) | LCCN 2017037501 (ebook)
ISBN 9780822372608 (ebook)
ISBN 9780822368601 (hardcover : alk. paper)
ISBN 9780822368724 (pbk. : alk. paper)
Subjects: LCSH: Epigenetics—Social aspects. |
Epigenetics—Philosophy | Art and science.
Classification: LCC QH450 (ebook) | LCC QH450 .S68 2017
(print) | DDC 572.8/65—dc23
LC recordavailableat https:// lccn.loc.gov/2017019584

Cover art: Anuradha Mathur and Dilip da Cunha,
Mumbai in an Estuary, from their *SOAK* exhibition
and book, 2009. Courtesy of the artists.

To Gowen Roper, always

CONTENTS

ACKNOWLEDGMENTS

My thanks go out to the many people, institutions, and programs that helped me as I was writing this book. Bruce Clarke jump-started one chapter by inviting me to the conference in honor of Lynn Margulis, where Susan Oyama was a wonderful interlocutor. Rachel C. Lee encouraged me to explore the feminist implications of this work at the Life [Un]Ltd. conference. The Gizmo Conference organized by Michael Bérubé and the Institute for Arts and Humanities at Penn State and the "Taking Animals Apart" conference at the University of Wisconsin, Madison, were both welcome chances to inhabit the contradictions of this subject. Particular thanks to the woman in the audience in Madison who suggested that I focus on the epigenetic landscape—you know who you are. In the United Kingdom, Sarah Franklin, Nick Hopwood, and Martin Johnson of Cambridge ReproSoc got me thinking about graphic embryos. Many thanks indeed to Sigrid Weigel, Vanessa Lux, and the Zentrum für Literatur- und Kulturforschung, Berlin, and to Lorraine Daston, Ohad Parnes, and the Max Planck Institute for the History of Science, for intellectual nourishment and challenge. To Art Laboratory Berlin, and especially to Regine Rapp and Christian de Lutz, I am so grateful for your warm welcome to a lively and inspiring interdisciplinary community. I thank Greg and Natascha for prompting us to go to Berlin in the first place, and our little Berlin community—Janina, Ohad, Sarah, Elliott, Irmela, and Stef—for making us welcome over many visits. My thanks to the fabulous Graphic Medicine community, especially Ian, MK, Michael, and Shelley; to our Penn State friends who have tolerated our jet-lagged reentry, especially Janet Lyon and Michael Bérubé; and to my beloved sisters Virginia and

Robin Squier, who not only came to Berlin to give us a chance to play but indulged our absence from the United States during a difficult year. Ken Wissoker has been an encouraging and patient editor, a rare and treasured combination. Many thanks to Anuradha Mathur and Dilip da Cunha for an engrossing interview and extended exchanges about their work; to Charles Jencks for his hospitality in welcoming me into his home for an afternoon interview; and to Anne Fausto-Sterling for stimulating Skype conversations. I thank Jenell Johnson and Irina Aristarkhova for their perceptive comments on several chapters; the errors that remain are definitely my own. Finally, I thank my graduate students Sara Di Caglio, Bethany Doane, Michelle Huang, Derek Lee, and Krista Quesenberry for their smarts and their sociability and the members of my graduate seminars on Graphic Medicine, Comics, Feminist Theory of the '60s and '70s, and Gender and Science: Reproduction. You taught me so much.

Last, I thank my children. Caitlin Squier-Roper's insights as an artist and landscape architect were a foundational inspiration for this book. I am very grateful to Tobias Gowen Squier-Roper for his scientific smarts and willingness to engage in long biological and medical conversations that were highly enlightening and great fun.

We have a joke in our house: Gowen has not read any of my books because he has *heard* them, endlessly, as I am writing them. But this time, he not only heard *Epigenetic Landscapes*; he read it. Every word. I am very grateful.

Figuring Development beyond the Gene

Figuring is a way of thinking or cogitating or meditating or hanging out with ideas. I'm interested in how figures help us avoid the deadly fantasy of the literal. Of course, the literal is another trope but we're going to hold the literal still for a minute, as the trope of no trope. Figures help us avoid the fantasy of "the one true meaning." They are simultaneously visual and narrative as well as mathematical. They are very sensual.

—DONNA HARAWAY, "Anthropocene, Capitalocene, Chthulhocene"

Epigenetics, "the study of changes in organisms caused by modification of gene expression rather than alteration of the genetic code itself," has been hailed in the popular press as a breakthrough field that can liberate us from the idea that we are controlled by our DNA.[1] Although the term "epigenetics" has been around in its current form since geneticist Conrad Hal Waddington introduced it in 1940, interest in this scientific field has spiked dramatically in the past several decades. Scholarly books on the topic proliferate.[2] The field appeals to so many because it seems to have a wide range of potential applications. To researchers interested in social, racial, and gender justice, the epigenetic dimension seems to hold exciting promise to free us from the idea that we are what our genes make us and enable us instead to identify those factors beyond genetics that shape us to become who we are. Could maternal diet, parenting style, or environment explain "the developmental origins of health and disease" (Loi et al. 2013, 142)? Did our grandmothers face starvation during pregnancy, leaving us a legacy of weight problems or undernourishment? Did a toxic physical or social environment limit our lung capacity or stress us so that we became vulnerable to depression? Epigenetics seems to reach from the body to society, holding out hope to illuminate issues as diverse as the development of gender identity; the intergenerational impact of slavery, war, or starvation; the range of factors that make us more vulnerable to depression or psychosis; or even the many variables that shape the health or illness of an ecosystem and the human beings dwelling within it (Fausto-Sterling 2012; Jablonka et al. 2014; Landecker 2011; Loi et al. 2013; Sullivan 2013).

These hyperbolic hopes may obscure the reality: there is significant uncertainty about the field of epigenetics. As this book was going to press, a widely cited study in *PloS Genetics* argued that epigenome-wide associations studies (EWAS) that claimed to document environmental contributions to heritable changes in disease risk were impossible to evaluate because of flaws in both their design and their execution. It was impossible, the researchers argued, to confirm "that epigenetics is responsible for the effects" the studies purport to show, and they concluded that "no EWAS to date can be said to be fully interpretable" (Birney et al. 2016). John M. Greally, an epigenetics researcher at Albert Einstein College of Medicine in New York City and one of the study's co-authors, explained to a *New York Times* reporter that a serious reexamination of the field was needed—*after* the team applied the classic remedy for research disappointment: "We need to get drunk, go home, have a bit of a cry, and then do something about it tomorrow" (quoted in Zimmer 2016).

In this study, I move away from the contemporary debates about epigenetics to focus on a figure that may help us understand the field afresh, a figure central to the development of this scientific field: the "epigenetic landscape." This book follows the cultural trail of the epigenetic landscape, a visual image developed by Waddington as the central figure for the scientific field of epigenetics, "the causal analysis of development" (Waddington 1940). As a scientific model, the epigenetic landscape fell out of use in the late 1960s, returning only with the advent of big-data genomic research in the twenty-first century; however, the figure of the epigenetic landscape is now being used across the life sciences because it enables scientists to think about, visualize, and communicate across disciplines and model development creatively.

Moving from the first version of the image—a landscape drawing created by the modern artist John Piper—to its later, more schematic versions, this book explores what the artistic and design elements of the image contributed to the meanings it held during the lifetime of its creator. Exploring the vital role the epigenetic landscape plays in fields beyond the life sciences, this study reveals that it has been used to model the intersecting complex systems that link scientific and cultural practices or, more precisely, reveal them as never having been separate or distinct. By examining three cases of such use—in graphic medicine, landscape architecture, and bioArt—this study reclaims the broader significance of this figure formed at the nexus of art, design, and science. It challenges the reductive understanding of epigenetics and argues instead for a more complex and varied view of biological development at all scales.

Waddington chose this visual image—in its first version, a charcoal drawing of a riverine landscape by Piper, and in its second and third iterations, schematic images of a ball on a contoured hillside—as a conceptual and methodological resource for those engaged in the "causal analysis of development" (Waddington 1940, 1). If we trace the significant aspects of this visual image through several different cultural realms, we can discover the broader conceptual and practical territories that are available to us when we explore development beyond the gene. I take a feminist science studies approach to my subject, inspired by Donna Haraway's "ongoing process of refiguring what counts as nature" and her commitment to escaping the "deadly fantasy of the literal," the "fantasy of 'the one true meaning'" (Haraway in Davis and Turpin 2015, 257). I hope to serve the same ends, by exploring the multiple meanings that epigenetics can hold for us as a field that is simultaneously visual and narrative, mathematical and sensual. By engaging with the figure at its center, I want to offer a new perspective on the scientific field of epigenetics and demonstrate that its complex, multidisciplinary origins have significant implications for the ways we understand and work with development more broadly.

The concept of epigenetics was formulated by the British geneticist Conrad Hal Waddington (1905–75), the son of the Quaker first cousins Hal Waddington and Mary Ellen Warner. His story is a striking mix of orthodox patriarchal British upbringing and maverick intellectual and social daring, and it will be helpful to have this in mind as we navigate the terrain of the epigenetic landscape.[3] Waddington saw his parents infrequently during his childhood; from his fourth year, they lived in southern India as tea planters while he was raised by an aunt and uncle back in England. As a child and young man, Waddington had very wide-ranging interests; inspired by relatives and friends with scientific and botanical interests, his passions ranged from naturalism and fossil collecting to visual art, poetry, and philosophy. At Cambridge University he studied paleontology and philosophy, writing his thesis on the mechanist-vitalist controversy. At and after Cambridge, Waddington demonstrated an interdisciplinary, synthetic approach to knowledge nourished by the intellectual catholicity of a good friend, the anthropologist, semiotician, and cyberneticist Gregory Bateson, as well as the artistic and design interests of the women he married. His first marriage took place during his Cambridge years, to a woman named Lascelles; when that marriage ended in 1936, he wed the architect Justin Blanco White. A close friend remembers evening discussions at the Waddington home in Cambridge that "used to cover not only science, but philosophy, modern art, music and the Dance." Waddington was an avid Morris dancer and an expert "exponent of its techniques" (Robertson 1977, 577). These marriages produced three children: Jake (later a physics professor), Caroline (a social anthropologist), and Dusa (a mathematician specializing in symplectic geometry and topology).

In his post-Cambridge years Waddington moved from paleontology to embryology and the investigation of biological development, working at the Strangeways Research Laboratory under Honor Fell.[4] Later, he served in the Operations Research Section of the Royal Air Force, supervising photoreconnaissance during World War II, before turning his attention in the postwar period to problems of population biology and animal genetics. He moved to Edinburgh in 1946 to head up the genetics section of the Institute of Animal Genetics. Waddington's fiftieth birthday was celebrated in 1955 with songs and poems, including one composed by the communist epidemiologist and geneticist Barnet Woolf, who also contributed music to works in the precursor of the Edinburgh fringe festival.[5] In his "Magic Words," a chorus of men explained epigenetics to the assembled celebrants (though they were probably already in the know):

If you want the correct explanation
Why embryos grow into men
The Alsatian begets an Alsatian
A hen's egg gives rise to a hen
Why insects result from pupation
Why poppies grow out of a seed
Then just murmur 'canalization'
For that is the word that you need.

Chorus   Then three cheers for canalization
Oh, come on now, hip hip hooray
A stiff dose of canalization
Will drive all your troubles away.
(Robertson 1977, 582–83)

The birthday gathering even featured "an epigenetic landscape constructed as a pinball machine, with the ball mostly travelling down the
main valley to produce normal phenotypes but on occasions being di-
verted into a secondary valley and producing a mutant" (Robertson
1977, 583; see also Goldberg et al. 2007). Waddington's birthday party
offers a whimsical glimpse of the variety of subjects before us in this
study: canalization and hen's eggs, entomology, botany, and embryology. We will come to all of these aspects of epigenetics, as well as to poetry, song, and the chance-laden factors that direct development into
one or another valley, whether microscopic or macroscopic, metaphoric
or material.

Ironically, what should have been the culmination of Waddington's
career, the creation of an Epigenetics Laboratory and an Epigenetics Research Group at Edinburgh of which Waddington would be "honorary director," was disappointingly derailed by the discovery of techniques for
hybridizing DNA and RNA. This decisively redirected scientific inquiry
from the study of development to the growing field of molecular biology. Waddington ended his career as an Albert Einstein Chair in Science
during a two-year stint as visiting scholar at the State University of New
York, Buffalo, where he taught a course titled "The Man-Made Future"
(Robertson 1977, 584). His daughter Dusa testified to that broad vision
of his later years, remembering her father in her Satter Prize acceptance
speech as "a Professor of Genetics who travelled all over the world and

wrote books on philosophy and art as well as developmental biology and the uses of technology."[6]

"Epigenetics" is a portmanteau word; Humpty Dumpty introduces this concept to Alice in *Through the Looking Glass* when she seeks his help with some words she cannot understand: "You see it's like a portmanteau—there are two meanings packed up into one word."[7] Perhaps Waddington paid homage to Lewis Carroll when he coined the term in the 1930s, for Carroll was among his good friend Gregory Bateson's favorite references.[8] The two meanings packed into this neologism fused the old Aristotelian expression for emergence, "epigenesis," with the rising field of genetics.[9] Waddington formulated this new field in his *Organisers and Genes* (1940). It would offer an analytic approach to development rather than the taxonomical and descriptive approach of embryologists up to that time. It was also in this work that Waddington presented the first version of his epigenetic landscape, a visual metaphor for the role played by stable pathways (later to be called "chreods") in the process of development. He elaborated on this theory in his later *The Strategy of the Genes* (1957).

The epigenetic landscape had only a brief heyday in its first run as a valuable scientific model. By 1961, when François Jacob and Jacques Monod discovered the *lac operon* (the combination of different genes involved in the metabolism of lactose), it was fading from use, seeming far too analogue and ambiguous to model processes that were increasingly capable of precise description (Baedke 2013; Gilbert 1991; Grene and Depew 2004).[10] Yet this changed when, in 2003, scientists and government officials at the National Human Genome Research Institute of the National Institutes of Health announced the completion of the Human Genome Project, asserting, "In addition to introducing large-scale approaches to biology, [it] has produced all sorts of new tools and technologies."[11] The challenges posed by "whole-genome" technologies, from biobanks and human genome databases to high-throughput screening techniques, catalyzed a return to the epigenetic landscape because it had the capacity to model the probabilities of change on a large scale.

Yet as the epigenetic landscape has come back into widespread use, it has done so with a difference. Now its scientific importance lies not in its representation of Waddington's "conceptual legacy," which was frequently overlooked in the rush to a molecular scale, but rather as a set of heuristic and methodological prompts (Baedke 2013, 756). The philosopher of science Jan Baedke has argued that during Waddington's research practice, the epigenetic landscape functioned heuristically in four ways: as visualization tools, as strategies for communicating across disciplines, as creative stimulation, and as methodological and modeling guides. In the post-Waddington era, Baedke argues, beyond the epigenetic landscape's primary utility for visualization, these same basic functions have continued in a wide range of fields across the life sciences, reaching even into the human and social sciences.[12] The epigenetic landscape functions in each case as a "tool" to "support transdisciplinary research; . . . stimulate visual thought [; and] guide modeling efforts and theory formation" (756).

In this book, I argue that the role of the epigenetic landscape extends beyond the life sciences. The inherent ambiguity of the epigenetic landscape as a metaphor gives it the potential to be a more productive model for the intersecting complex systems that are now understood to link scientific and cultural practices—or, more precisely, to reveal them as never having been separate or distinct. Historians, philosophers, and rhetoricians of science have demonstrated that metaphor plays an important epistemological and rhetorical role in scientific thought, for good and ill, by transferring meaning from one context to another; preserving aspects of previous thought styles in new areas; consolidating or disrupting gender relations; catalyzing experiments; and generating new frames for thinking, reading, and writing, as well as foreclosing others. Models, too, exist in the liminal zone between scientific theory building and scientific practice, where they can induce thought, touch, and movement to draw the model user into a conceptual space and engage her with its questions (Keller 2000; Myers 2015). Indeed, Evelyn Fox Keller (2000, S77) argues that "metaphors, like models (indeed a crucial component of many models), can themselves function as tools for material innovation."[13] I track some of these innovations in the body of this book.

My aim in what follows is to recover the expansive reach that epigenetics had as Waddington worked with the term over the course of his life. With his conceptual legacy obscured, the meaning of the term "epigenetics" since his death has been increasingly focused—indeed, it has been narrowed. Now the term refers primarily to the specific mechanisms by which epigenetics works on a molecular level, particularly DNA methylation and chromatin modification (Feinberg 2008, 1345; Jablonka et al. 2014, 393). When the term "epigenetics" appears in research exploring development in fields as widespread as oncology, environmental toxicology, prenatal medicine, nutrition, and psychiatry, its meaning usually tilts away from the macro meaning it once had and toward the micro realm. Contemporary epigenetics research frequently affirms a linear, gene-centered, and "programmed" approach to development, a kind of "somatic determinism," according to the historian of science Sarah S. Richardson (2015, 217, citing Locke 2013, 1896).[14]

This contemporary narrowing of epigenetics has already inspired critiques of the field from a feminist perspective. Let me give some examples. Findings about the role of epigenetics in sex and gender differences are often interpreted to endorse existing conceptions of sex and gender as binary, programmable, and stably retained over time (Richardson 2015). Epigenetic research is being directed—one could even say contained—to the kinds of studies that can add new tools or explanations to our existing framework for understanding development (Richardson, forthcoming). Similarly, some scientists working in evolutionary-developmental ecological biology (evo-devo-eco) or eco-devo-evo (the order of the abbreviations packs a partisan punch) are turning to epigenetics because they hope its study will enable them to identify how an organism integrates its genetic, environmental, and developmental processes, information they then plan to integrate into existing evolutionary theory (Abouheif et al. 2014). Even when researchers do seem willing to confirm the paradigm-shattering implications of epigenetics, acknowledging the complex and nonlinear view of development their research has revealed, they may actually be engaged in a rhetorical holding action, trying to reorient the program of epigenetics research toward their specialty to cope with increasingly scarce resources and rising demands for concrete results (Panovsky 2015, citing Arribas-Ayllon et al. 2010).

There has been a tension in the understanding of epigenetics: should it be framed narrowly or broadly? In terms of gene action or of developmental plasticity? Although the dominant strategy of the field has been to use epigenetic findings to support the mainstream gene-centered view, the feminist and postgenomic critique cited above reflects the view of other researchers that epigenetics could support a new understanding of biological organization that stresses plasticity rather than genetic determinism (Love 2010; Shapiro 2015; Sullivan 2013; Van Speybroeck 2002, 61, 79).

This tension flared into a firestorm in the response to an article published in the *New Yorker* magazine in May 2016 by the cancer researcher Siddhartha Mukherjee. In a Lewis Carroll–like bit of wordplay, Mukherjee's essay, "Same but Different," used his twin aunts to illustrate the impact of environmental factors in development, drawing parallels to the social behavior of the jumping ants studied in the New York University School of Medicine laboratory of the epigeneticist Danny Reinberg (Mukherjee 2016b). When Mukherjee's article appeared in print, the forces of disciplinary normalization came out in force, with more than one hundred scientists issuing accusations that he overemphasized epigenetic mechanisms and neglected the role of genetics (Mukherjee 2016b). As Chris Woolston reported in *Nature*, Mukherjee acknowledged his mistake: "He put too much emphasis on the 'speculative roles' of histone modification and DNA methylation. 'This was an error,' he says, adding that a mention of transcription factors could have helped to avoid 'an unnecessarily polarizing reading of the piece,'" (Woolston 2016, 295).

Woolston also described the attempt by John Greally to put the episode in context: "Greally adds that it's hard for anyone to talk about epigenetics without stirring up controversy. Different researchers have different definitions for the term, and there are still many questions about the mechanisms behind the regulation of gene expression. 'We're in a bit of a mess in epigenetics,' Greally says. Mukherjee is 'a thoughtful guy,' he adds. 'But he's beginning to realize that he stepped on a land mine.'" A land mine, indeed. The very next month, in June 2016, Greally and his fellow researchers published their article in *PLoS Genetics* sounding the alarm about EWAS and calling for a major reassessment of epigenetic research. Earlier in this chapter I quoted the interview in which he jokingly suggested he had been driven to drink by the methodological problems he found in those studies. Yet the *New Yorker* may have helped Greally pack the explosives into that land mine by choosing as the subtitle to Mukherjee's article "How Epigenetics Can Blur the Line between Nature and Nurture."[15]

The phrase seems to allude to an incisive and influential entry into the discussion of epigenetics, Keller's slim volume *The Mirage of a Space between Nature and Nurture* (2010).[16] There, Keller charges that the ambiguity, confusion, and general muddle in our understanding of nature and nurture can be attributed to shortcomings in the language of genetics. The problem is not only the discourse of "gene action," which Keller has so powerfully critiqued, but the "chronic slippage between the two meanings—ordinary and technical—of *heritability*" (or, to think of it in Mukherjee's framing of the problem, between his aunts and the jumping ants). "Heritability" is an ambiguous word, Keller points out, because the means of transmission of traits between the generations can be "genetic, epigenetic, cultural, or even linguistic." Writing of geneticists and molecular biologists, she observes, "When the words they use have multiple meanings, meaning is not so easy to control. . . . Consciously or not, slippage happens; it is not only easy to mean two—or even three—things at once, it may be unavoidable. What is difficult is meaning only one thing" (Keller 2010, 71). For Keller, it is not precisely the ambiguity that causes the problem but our failure to recognize it. "The problem is that, as the different meanings of the term travel back and forth between different kinds of arguments, different logics, and different disciplines," she writes, "the ensemble becomes knitted together into a seemingly coherent whole, giving rise to a seemingly coherent argument" (75–76). "Seemingly," here, is the keyword.

While linguistic ambiguity can be confounding or productive, depending on whether we are attentive to its presence or sink into the delusion that the word or phrase in question means "only one thing," the visual ambiguity of figures can be epistemologically enabling, complicating our thinking, encouraging us to cogitate, meditate, or just "[hang] out with ideas" and helping us to "avoid the fantasy of 'The One True Meaning'" (Haraway in Davis and Turpin 2015, 257; Keller 2010, 76). "Images function effectively at drawing viewers in, confounding them, and prodding them to ask questions" (Allen 2015, 141). Models provoke engagement, temping people to touch them, explore their surfaces, and even move with them (Myers 2015). Sensual, visual, narrative, and even mathematical, Haraway reminds us, figures can help us figure things out. Taking advantage of their epistemologically productive ambiguity, this study focuses on the three major visual images of the epigenetic landscape that Waddington used in his scientific publications between 1940 and 1957: "the river," "the ball on the hill," and "the view from underneath with guy wires."[17] In what follows, I explore the dramatic differences among the three versions in origin, subject matter, mode of composition, semiotics, and, most of all, the epistemology and ontology they imply.

The original epigenetic landscape was a work of landscape art, commissioned by Waddington from his friend John Piper. We will look much more closely at it in a later chapter, but for now I will just describe it briefly: in shades of gray, white, and black, it shows a turbulent river flowing through brush-bordered banks. As its caption as the frontispiece in *Organisers and Genes* reveals, this version of the epigenetic landscape also has an element of fantasy: "Looking down the main valley towards the sea. As the river flows away into the mountains it passes a hanging valley, and then two branch valleys, on its left bank. In the distance the sides of the valleys are steeper and more canyon-like." Despite the gravitational paradox—how can a river flow away into the mountains?—Waddington judges it "an amusing landscape to picture to oneself; and I think it expresses, formally at least, some characteristics of development which are not easy to grasp in any other way" (Waddington 1940, 93).[18]

While the ball on the hill and the view from underneath with guy wires are quite different visually and compositionally from the first version, the second and third versions of the epigenetic landscape also express its three central principles: canalization, homeorhesis, and scaling. We will encounter these processes later in much more detail, but for now here is a brief definition of each. Canalization, or developmental robustness, is the ability to sustain a developmental direction despite environmental disruptions. Waddington termed these dedicated developmental pathways "chreodes" (another coinage). Homeorhesis, which is related to the physiological concept of homeostasis, or the maintenance of equilibrium, in contrast, is the ability of a dynamic system to sustain its rate of change or flow. Temporal and spatial scaling are the perceptual/conceptual properties that make this developmental model of the "biological picture" inclusive of life from conception to death and meaningful at the scale of a cell, an embryo, or a population.[19]

Figures are sensual, Haraway tells us. As a model that is also a metaphor, the epigenetic landscape engages our senses, as an example from Waddington's late-life writings can reveal. In his final, posthumously published book, *Tools for Thought: How to Understand and Apply the Latest Scientific Techniques of Problem Solving* (1977), Waddington included a set of instructions for "Exploring a Landscape." Inspired by a paper by the Russian mathematicians Israel Gel'fand and Michael Tsetlin (translated from the Russian by his daughter, the mathematician Dusa McDuff), Waddington imagined using kinesthetic strategies to investigate the unknown in the epigenetic landscape. "An important question about the epigenetic landscape and branching pathways is this: When we are confronted with an unknown system, how do we find out what the shape of the landscape is?" he writes. "One suggestion, due to . . . Gel'fand and Tsetlin, is to proceed as follows. We find ourselves doing something to a system which we believe has certain stability characteristics, which could be described as an epigenetic landscape; but we have no idea where we are on the landscape when we first start trying to affect the system" (Waddington 1977, 113).

The entire passage repays careful reading for its vigorous kinetic language. Waddington imagines himself "going out into the landscape," moving uphill and downhill, following the slope into the valley, and then taking "quite a large jump" across the landscape "onto the opposite hillside lower down the valley," where "a local exploration around that spot may show us the slope going in the opposite direction." The exploration works by trial and error. As he explains, "One can't, of course, give any general rules for doing this. It has got to be largely a method of 'suck it and see.' A point of general principle is that in exploring such a landscape it would take too long to walk all over it step by step. . . . It is better to alternate between (a) local exploration . . . and (b) a jump in the dark to try to change some quite different aspect of the system" (Waddington 1977, 113–14).

This combination of local exploration and the jump in the dark with which Waddington proposes to explore the epigenetic landscape is familiar to me from years of my own research. In an earlier book, I thought of it in terms of escaping the academic culture of expertise, giving myself the holiday of curiosity, or "poach[ing] on academic territory in which I can claim at best amateur competence."[20] Now I think of it as the strategy of refusing critique to follow concern instead (Latour 2004).

Waddington's instructions for "Exploring a Landscape" also bring to mind the philosopher Michel Serres's *tiers-instruit*: a third mode of learning that operates not through critique and subordination to one reigning epistemological category, but as a wandering, translational commentary that ranges across disciplines and disciplinary languages. Rather than dividing and subordinating fields, Serres multiplies them and disturbs their boundaries, preferring disorder and fertility to sterile order. In my thinking about the epigenetic landscape, I am also inspired by Serres's method of inquiry, which, as Bruno Latour has explained in a useful pair of images, differs from the standard Western epistemological model (Latour 1987, Serres and Latour 1995). Because one of the main points of this book is the importance of attending to the productive expressiveness of figures, I include them here (figure I.1).

The first image, a circle marked by arrows extending both outward toward the periphery and inward through the "intermediary" to the center, represents "a powerful critique . . . that ties, like a bicycle wheel, every point of a periphery to one term of the centre through the intermediary of a proxy" (Latour 1987, 90). Latour describes this as the mode of the "Critique philosophers," who "firmly install their metalanguage in the center, and slowly *substitute* their arguments to every single object of the periphery" (90). The second image represents Serres's "pre-critical philosophy," a series of parallel lines stacked one above the other, with the tiered labels "Language 1, 1.2, 1.3, and 1.4" and wavering and straight vertical lines linking the tiers. "Crossover from one repertoire to another," the caption reads. Latour defines Serres's method not as critique but as commentary, a "cross-over, in the genetic sense, whereby characters of one language are crossed with attributes of another origin" (90–91).

Beneath the linguistic layering, there is something topographic in this image, an anticipation of the contour grooves we will encounter in the second image of the epigenetic landscape. The negotiation of the tiers in Latour's second image (a process both linguistic and spatial) resembles the challenge posed by another medium to embody the perspective of

Periphery

Center

Intermediary

Language 1

1.2

1.3

1.4

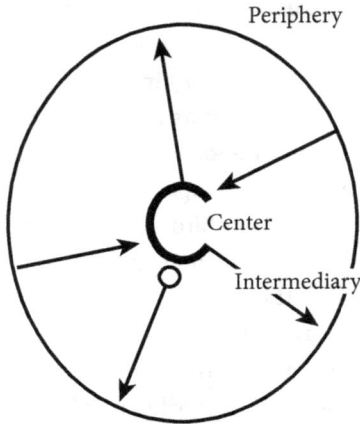

Substitution of the metalanguage
to the infralanguages of the periphery

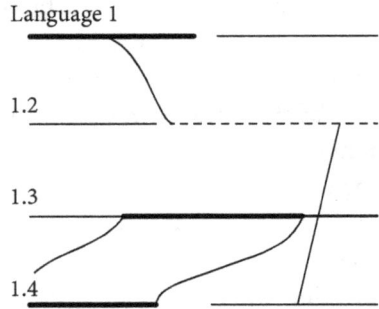

Crossover from one
repertoire to another

FIGURE I.1 Michel Serres's method of inquiry, as imaged by Bruno Latour (1987, 90).

the epigenetic landscape. As we will see in a later chapter, the medium of comics requires the reader/viewer to follow a verbal and visual narrative across panels, down tiers, and through gutters while continuously supplying the closure the narrative requires. While Latour's celebration of Serres emphasizes his understanding of the relations among language, biology, information theory, and thermodynamics—a view that arguably is heavily indebted to Waddington—it also illuminates the methodological choices I faced while writing this book. I could have situated my discussion of epigenetics within the broad frames of developmental systems biology, regenerative medicine, postgenomics, or bioethics, choosing one of them as the centering strategy within which to consider the epigenetic landscape as a metaphor and model. Instead, in accord with what Rick Dolphijn and Iris van der Tuin have dubbed "the transversality of new materialism . . . a nomadic traversing of the territories of science *and* the humanities, that perform[s] the agential or *non-innocent* nature of all matter," I investigate the relations between objects and subjects, states and forces customarily held in opposition (Dolphijn and van der Tuin 2012, 100–101; see also Braidotti 2006). Or, to put it more simply, I have followed my developing thoughts where they led me, even (especially) if they drew me laterally rather than linearly. After all, despite disciplinarily formed practices of denial, scientists, writers, and artists have long been thinking, learning, teaching, and inquiring collaboratively. As feminist science studies has shown, biology has been shaped all along by both aesthetic and social concerns, just as the humanities and arts have engaged with the vital process of development (Haraway 1976; Haraway in Davis and Turpin, 2015).[21] I have chosen to combine local exploration with many a jump in the dark as I investigate how the epigenetic landscape can illuminate our understanding of disciplines marginal to the life sciences and even challenge our habitus toward disciplinarity.

So much is at stake in working with the epigenetic landscape that my investigation is unavoidably incomplete.[22] While exploring the creative potential of the epigenetic landscape as an instructive third space, I have tried to stay close to the method Waddington established so I can recapture the broader implications contained not only in the epigenetic landscape, but also in the concept of epigenetics. Waddington hoped for a crossing over between embryology (which studied the impact of surgical or chemical interventions in a living embryo) and genetics (which studied the role of hereditary factors, later known as genes, in producing embryonic changes). I hope through my own post-disciplinary practice of crossing over to show how the scientific study of development can illuminate ways to work with developing life beyond those initial focuses.

Although I have restricted my attention to areas suggested by the specific images provided by the epigenetic landscape itself—the ball on the hill, the river, and the view from underneath with guy wires—I explore what each image affords as a means of orientation as I carry out a local exploration of the specific environment to which it has drawn me. As a result, while this project has been a bit of a jump in the dark, it has taken me deeply and pleasurably into very different fields of endeavor. In each field I have chosen one version of the visual image of the epigenetic landscape, the one that originally brought the field to my mind, as my prompt or tool as I explore the strategies each field uses, though at very different scales, to work with the balance of freedom and constraint, change and stability, inherent in biological development.

The ability to think about development from a number of different perspectives, visual as well as verbal, fuzzy as well as precise, is increasingly understood to be a catalyst to creativity (Meloni and Testa 2014). Such a process of exploring the unknown is even more pertinent to the power of the epigenetic landscape. This visual image was not only attractive, alluring, contradictory, and even seductive but also deeply playful, as Waddington's set of instructions for exploring it reveal. While we have learned in science studies to appreciate the epistemological values of the imprecise, the allure of the unknown "epistemic thing" that tempts us out of our comfort zone, and the challenge of forging an "epistemology of the concrete," we are only beginning to push these strategies to their limit by watching them as they play, as well as play out, beyond the realm of science (Rheinberger 2010).

Therefore, I approach these versions of the epigenetic landscape as models not merely in the sense of being predictable or testable scientific objects, but also—in Waddington's tradition—as productive engagements with the unknown. I understand them to function kinetically, affectively, and methodologically, as well as epistemologically. Just as the embryologist Wilhelm His found in the act of creating the wooden model of an embryo the capacity to integrate perceptions previously held at a distance, and just as protein crystallographers dance into the proteins whose folding they are attempting to model, so, too, the enactments of the epigenetic landscape we will look at in the chapters that follow provide opportunities to make development physiologically concrete, affectively present, and methodologically meaningful (Hopwood 1999; Myers 2015).[23]

The first two chapters of this study follow the epigenetic landscape from its origins as a nexus of Waddington's scientific and artistic interests, through the different functions it held for him in the 1960s, and finally to its methodological expansion in the work he did in his final years. I draw on the concept of canalization as I explore how the gendered and disciplining effects of different intellectual and social contexts shaped the use and reception of the epigenetic landscape. While interest in the epigenetic landscape waned during the Serbelloni Symposia, which Waddington convened to frame the discipline of theoretical biology, it returned and expanded while he was writing *Behind Appearance: A Study of the Relations between Painting and the Natural Sciences in This Century* (1970 [1969]) and still more in his late-life projects *Tools for Thought* and, with Erich Jantsch, *Evolution and Consciousness: Human Systems in Transition* (1976).

Each of the following chapters moves beyond Waddington to take one instantiation of the epigenetic landscape—the ball on the hill, the river, and the view from underneath with guy wires—as its methodological model. Working from the central image to the specific field of practice it has suggested to me, I explore what this version of the epigenetic landscape brings to that local system of thought and practice as it deals with development. Chapters 3 and 4 focus on the most familiar image of the epigenetic landscape, the ball or fertilized egg or embryo poised near the top of a contour-riven slope. Because this image suggests the field of embryology, and particularly the history of making embryos visible, I consider the temporally and spatially scaled nature—the *Russian dollness*, if you will—of developmental processes in chapter 3. I explore how the shift from descriptive and taxonomic embryology to analytic embryology not only reflected but was enabled by what Janina Wellmann (2017, 31) calls a new "epistemology of rhythm." Expressed across the entire cultural field through a mode of serial graphic display, this strategy of combining movement and stillness, image and gap, not only nurtured a newly process oriented perspective but also prepared the way for the stop-motion photography and animation that would later bring embryology to widespread public attention.

In chapter 4, I continue this analysis, turning to the nexus of contemporary popular culture and medical communication, where the comics medium provides a powerful space in which to image and enact the process of development at multiple scales. In their temporal and spatial complexity, comics recover the properties of the early twentieth-century embryo cartoons and animations that made it possible for embryologists not only to analyze development but also to share their findings widely. I describe how graphic medicine—comics about illness, medical treatment, disability, and caregiving—are providing a remedy for the narrow instrumentalism of the institution of medicine. Looking closely at what I call the graphic embryo, a comics genre grounded in the tradition of embryo imaging, I argue that as it remediates the medical image of the embryo, it provides an aesthetic and social space to reimagine development. By "unflattening" it into multidimensional time and space, the graphic embryo offers the possibility of nonlinear outcomes, diverse developmental trajectories, and a more complex model of development that includes not only embryos, but also infants, children, mothers, fathers and even a "sentient organism in [a] nearby solar system" (Sousanis 2015; Nilsen 2014).

Chapters 5 and 6 return to the first version of the epigenetic landscape, Piper's drawing of the river, and the particular epigenetic principles it expresses. Beginning with Waddington's late-life application of the epigenetic landscape model to ecology, chapter 5 traces the entangled commitments to ecological mapping and scientific control in the work of Waddington's fellow Scot Ian McHarg, founder of the field of landscape architecture. A look at several other prominent landscape architects working in McHarg's tradition reveals that the epigenetic principle of homeorhesis, the maintenance of a steady rate of flow or change, has been both harnessed and transgressed by landscape theorists as they deal with development on a macro scale. Chapter 6 continues this examination, moving from McHarg's mode of landscape architecture theory and practice; the contribution of feminist landscape architectural theory; and a more expansive understanding of development to, finally, the exciting contemporary work of Anuradha Mathur and Dilip da Cunha, who reorient the landscape architectural treatment of development from reductionist linearity to situated, kinetic complexity, with ecological and global sociopolitical significance.

Chapter 7 turns to the final version of the epigenetic landscape I explore in depth: the view from underneath. Originally designed to reveal the genes and guy wires whose interactions produce gene expression and thus shape the contours of the epigenetic landscape, this image served Waddington in his later life as a model of social and conceptual development and more recently has been adapted to represent complex sociological processes (Tavory et al. 2012, 2013). In the chapter, I profile Art Laboratory Berlin, an epigenetic landscape in its full transdisciplinary sense. I argue that in this space that entangles art and science, practitioners and members of the public model a highly speculative, kinetic, and affective multidisciplinary approach to producing knowledge about development.

The conclusion provides a brief sketch of how one contemporary feminist scientist is using cartoon animation to adapt the three images of the epigenetic landscape—the ball on the hill, the river, and the view from underneath with guy wires—to illuminate the development of gender identity and sex differences in behavior. Returning to how one life scientist is currently using the epigenetic landscape as an epistemological and

analytic prompt in her research, we appreciate her incorporation of seriality, rhythm, art, design and ecology, all aspects of the epigenetic landscape that have been not just maintained but also productively extended
by fields beyond the life sciences. Anne Fausto-Sterling's selection of cartoon images to figure, and figure out, the processes of gender role development illustrates the collaborative, transdisciplinary, and reflexive uses
of the epigenetic landscape that this study explores. Fausto-Sterling's adaptations of the epigenetic landscape complicate our understanding of
epigenetics, helping us to "avoid the fantasy of 'the one true meaning'"
and to see instead, with Lily Briscoe, that "nothing was simply one thing"
(Haraway in Davis and Turpin, 2015; Woolf 2007, 376).

CHAPTER 1

⌄

The Epigenetic Landscape

The river cascades steeply down a deepening valley towards a high
horizon at the upper right of the picture plane. Just under the top
border of the picture, heavy-bellied clouds float above what might be
the ocean. In the lower right foreground what seem to be bramble
bushes reach almost to the edge of the roiling river. The left side of the
landscape seems strangely regular: three parallel deeply grooved side
channels split off in mirroring arcs to the left from the diagonal slash
of the river. Yet the caption of the drawing describes an impossible
geography in which water flows both up into the mountains and down
into the sea: "Looking down the main valley towards the sea. As the
river flows away into the mountains it passes a hanging valley, then
two branch valleys, on its left bank. In the distance the sides of the
valleys are steeper and more canyon-like."

—C. H. WADDINGTON, *Organisers and Genes*

This evocative work of art is the first version of the epigenetic landscape.
A drawing by the British artist John Piper that was commissioned by his
friend the embryologist C. H. Waddington, it appeared in 1940 as the frontispiece to Waddington's pivotal study of embryonic development, *Organisers and Genes*. Waddington hoped that this work would create a bridge
between embryology and genetics, so it seems likely that he intended the
frontispiece to prepare readers conceptually for this by illustrating the process by which a fertilized egg develops, as different parts of embryonic
tissue trigger the downstream development of other parts. Elsewhere in
the volume he relied on other sorts of diagrams, since for several years

he had experimented with adapting the branched-track diagrams used by the railway; the images that represent the directions a cell takes as it develops to differentiation (known as cell-fate diagrams); and fitness landscapes and adaptive landscapes, the images that use contrasting peaks and valleys in a metaphoric landmass to represent how advantageous or detrimental different groups of genes prove to be. (For a number of good discussions of these different models, see Baedke 2013; Gilbert 1991; Ruse 1990.) Yet curiously, the picture he chose for the frontispiece of this important volume is not schematic like the diagrams that follow in the text; nor does it picture a fertilized egg or embryo, whose developmental process provides the book with its main subject. Instead, it is a work of art: Piper's drawing of a river flowing between deeply fissured banks (figure 1.1).

Most discussions of the epigenetic landscape have tended to move on quickly from that anomalous first image to the familiar one that followed it seventeen years later, a schematic line drawing of a ball perched near the top of a scored incline. However, because I am interested throughout this book in recovering the broader significance of the epigenetic landscape, I want to begin by asking why Waddington chose to commission a work of art as its first representation. What was the context for that decision, and what can we learn from it about the way he framed the new field of epigenetics? I will argue that the movement from the original drawing of a landscape to the later schematic geometrical image of the embryo on a hill reveals a tension in epigenetics as a field that continues to this day. We could frame this tension in several ways: between the aesthetic and the programmatic, between context and content, or between holism and reductionism. I track this tension first as it figures in the changed visualization of the epigenetic landscape; next as it appears during a later period in Waddington's life, during which he was attempting to theorize both biology and art; and finally in his two posthumous volumes, where he extended his formulation of the epigenetic landscape beyond its applications in biology. The developing versions of the epigenetic landscape can also be read for the tension between situated knowledge and decontextualized abstraction, a feminist framing that recurs in the chapters that follow.

In this chapter, I work from the initial context within which Waddington first commissioned the visual image of the epigenetic landscape

FIGURE 1.1 John Piper's drawing of the river: The first version of the epigenetic landscape (Waddington 1940, frontispiece).

through its later revisions. I argue that the massive environmental disturbance of World War II and the Blitz stimulated Waddington's thinking, encouraging him to explore interests both scientific and artistic. In the postwar era and certainly by the 1950s and 1960s, Waddington's scientific and artistic interests became less mutually informing. Instead, they were in tension, reflecting the different disciplined and gendered contexts in which he was working. This tension came to shape how he approached epigenetics in general and the development of the epigenetic landscape specifically.[1] Drawing on these findings, I show how Waddington's work in the scientific context gradually drew the epigenetic landscape away from its artistic roots. Yet as I also show, aspects of the original, visual image of the epigenetic landscape persisted as robust models for understanding the implications of epigenetics for fields beyond the life sciences—specifically, the arts and humanities. This chapter prepares the way for the chapters to come, each of which focuses on one of the principal versions of the epigenetic landscape model in Waddington's work: the river, the embryo, and, finally, the perspective from beneath the EL, or, as Waddington captioned it, "The complex system of interactions underlying the epigenetic landscape," and its more recent elaborations. I explore the ways these three versions illuminate the transdisciplinary potential of this scientific model, the product of a collaboration among art, science, and technology.

PIPER AND WADDINGTON

When *Organisers and Genes* was written, Piper and Waddington were already friends. Indeed, Waddington acknowledged his friend's artistic contribution to the work in his preface: "I am also grateful to John Piper for his interpretation of my somewhat romantic conceit, the epigenetic landscape." In 1936, Waddington married the architect Justin Blanco White, a friend of Piper's wife, Myfanwy, and soon became an admirer of Piper's work. Then a fellow and lecturer in genetics at Cambridge, Waddington was living in rooms at Christ College, and it was there that he hung Piper's "Painting 1935." Waddington leased the large abstract work from Piper for three pounds after it appeared in the first exhibit of abstract art in England (Spalding 2009, 76).

Not only was art changing in the late 1930s and early 1940s, but that period marked an important shift in the careers of both Waddington and

Piper. Both men had been concerned in their respective fields with the problem of representing the impact of the passage of time, according to the historian of science Ohad Parnes.[2] For Piper, the challenge of painting temporal processes was an aesthetic one, a matter of the tension between abstract and figurative modes of painting, as well as an index of his increasing awareness of the connection between the movement of time and changes in the landscape. For Waddington, the challenge was epistemological: he wanted to find a way to link embryological development to hereditary transmission through the representation of temporal processes in a visual form. For both of them, a turn to topography in the first model of the epigenetic landscape provided the solution, retaining both aesthetics and epistemology. Although Waddington probably commissioned Piper's drawing in 1939, it appeared first in print as the frontispiece of *Organisers and Genes* in 1940, nearly simultaneously with the beginning of the London Blitz. Given its wartime context, the epigenetic landscape was charged with meaning for both men, although the meanings it held reflected their different contexts.

The very recourse to landscape imagery may be surprising to those who remember Piper as he was first widely known: as an abstract artist, passionate in defense of nonrepresentational painting. Yet over the course of his career, Piper became one of the most important and beloved of Britain's landscape painters. By the end of the 1930s, despite his success in abstract art, Piper had begun to lean away from abstraction and toward the figurative.[3] Several experiences, Parnes suggests, fueled a new conviction that landscape painting posed the greatest challenge to the artist of his era. In "Lost, A Valuable Object" (Piper 1937), an essay that reads now as an early foray into object-oriented ontology and a patriotic call to painterly arms, Piper made a plea to artists to return to painting the land in all of its material reality. Neither surrealists nor abstract painters deign to acknowledge the object "*in its proper context,*" he complained (his italics). "The one thing neither of them would dream of painting is a tree standing in the field. For the tree standing in the field has practically no meaning at all for the painter. It is an ideal: not a reality" (70). The essay was written when Germany had invaded the Rhineland and Francisco Franco was the head of the Spanish state, and Piper clearly felt that the country was on the verge of war. We can hear a distinctly political and nationalist cast to his argument that "the object must grow again; must reappear as the 'country' that inspires painting. (It may, at the worst, turn out to be a night-bomber, or reappear in a birth-control poster—but it will grow again, somehow.)" (72).

Piper had another reason to be thinking about the British landscape in the late 1930s. He had been working as a topographer with Osbert Guy Stanhope Crawford, a geographer-turned-archaeologist who relied on aerial photographs of the landscape to determine the location of archaeological sites. This skill was useful when Britain declared war on Germany in September 1939 and Piper became an official war artist. He was initially sent to paint Coventry Cathedral when it was bombed during the Coventry Blitz of November 1940, and later he began to paint buildings smashed or threatened by bombing, ultimately producing the series of works that would make his fame. As his biographer Frances Spaulding quipped, "Scenes of devastation made John a household name" (Parnes 2007, 2015).

As Piper was revising his approach to painting, Waddington was engaged in a parallel rethinking of representation. With the onset of World War II, Waddington returned from the genetics laboratory at the California Institute of Technology, where he had been studying the wing development of fruit flies (*Drosophila*), to the British halls of military advisers. He became engaged in wartime service, first working for the Royal Air Force in operations research tracking U-boats, and later serving as scientific adviser to the commander-in-chief of the British Coastal Command.[4] Probably influenced by Piper's return to landscape painting and their collaboration on the model of the epigenetic landscape, in 1941 he published "Art between the Wars," an essay in which he discusses how scientists and artists should respond to the wartime experience. He echoes Piper's critique that modern art is failing to "get ahold of real things" (Waddington 1941, 47). Yet tellingly, Waddington describes those "real things" as requiring not only the figurative aesthetic for which Piper called, but also the scientific reasoning central to his own epistemology.

"Art between the Wars" opens with a topographical metaphor that not only reveals his debt to Piper but also demonstrates Waddington's keen awareness of the gradations of literary and geographical hierarchies. It situates poetry and painting in human culture represented as a landscape: "If one compares a whole culture to a valley, the novels are the great rivers of the plain, on which the traffic flows; but the same slope of the land is more obvious up in the uncouth hills where the little streams, the poetry and painting, make a great clatter but gather no moss." Dismissing the poetry of T. S. Eliot and e. e. cummings because of its "emphasis on the base and ignoble results of our social existence," the essay moves on to modern art. Praising Piper's critique of modern painters in "Lost, A Valuable Object" for failing "to find any real tree in the field," Waddington applauds the fact that painters have been forced by the economic stringencies of wartime to return to figurative painting. While he acknowledges that the only painters able to keep on painting have been required to work as war artists, documenting the disorientation of battle scenes or providing visual records of the threatened "monuments of England," he praises Piper and his contemporary war artists Henry Moore, Graham Sutherland, and Paul Nash for producing "first-class art and first-class reporting." Not only are they rescuing art from the curse of specialist aesthetic technicalities, Waddington opines, they are also giving it a secure place in the British cultural landscape, redefining art as "once again an important element in the main stream of civilization" (Parnes 2007, 9; Waddington 1941, 37, 40, 48–49).

World War II was the context that shaped the first version of the epigenetic landscape as a visual model for development. As the painter and the embryologist-turning-geneticist were both confronted by the importance of landscape, they came to see it in the broader wartime context. The landscape around them in the late 1930s and early 1940s bore marks of the passage of time at multiple scales, as the impending war threatened individuals and populations. It was not just simple topography but a source of valuable memories and hopeful possibilities. It is initially surprising, then, that the first image of the epigenetic landscape registers none of these human impacts of war. The turbulent river flows between unpeopled banks, and only the brambles on its right bank and the impossibly dual direction of its flow attest to the difficult circumstances within which Piper and Waddington collaborated to create this visual image.

When the next drawing of the epigenetic landscape appeared in print, seventeen years later, it offered a dramatically different kind of visual model for representing development. Gone was the situated scenic realism of Piper's drawing, with the suggestively moody naturalist aesthetic of landscape art. Instead, *The Strategy of the Genes* (1957) pictured a highly schematic image: a ball perched near the top of a hill fissured with channels indicated by contoured lines, some branching off, some not, and

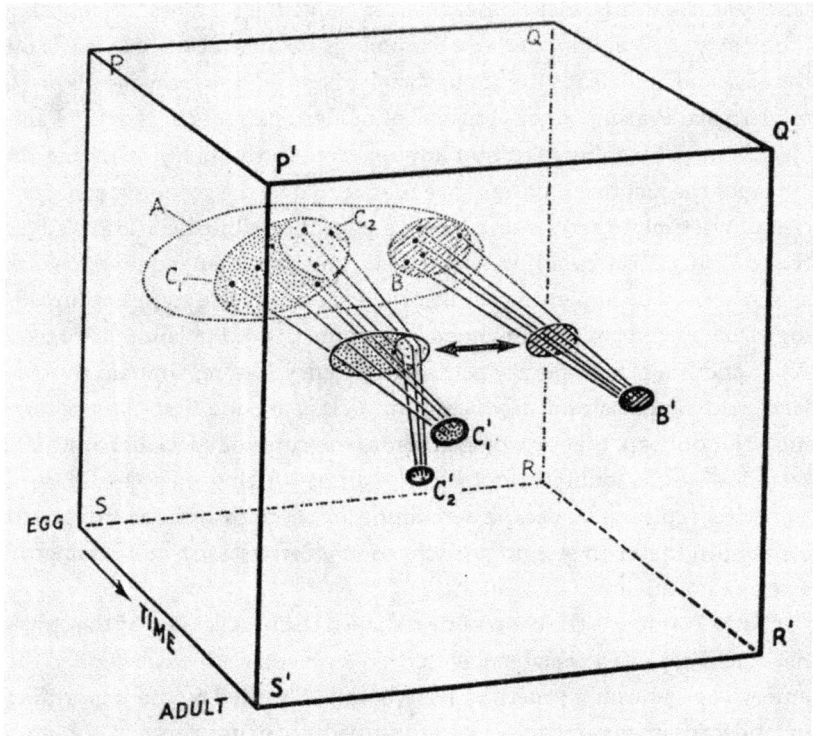

FIGURE 1.2 "A phase-space diagram of development," Waddington's discarded alternative for the epigenetic landscape (Waddington 1957, loc. 566).

most of them finally ending up, one way or another, at the bottom of the hill. The only remnant of the Piper-Waddington collaboration appears in those channels, which are represented by finely drawn parallel lines reminiscent of the contour maps of Waddington's early days as a geologist and Piper's work as a topographer.

What could this drastic change in the epigenetic landscape mean? Waddington's impulses had been shifting since the first version, to be sure. In the intervening years, he had toyed with the idea of a geometrical model represented in multidimensional phase space (figure 1.2)—that is, one that represented all of the possible outcomes of the changing system, since he believed that the multiple interactions and probabilities of epigenetics could not be represented accurately in a two-dimensional drawing.

He ultimately discarded this image, opting instead for the image of a ball on a hill because he felt it was more accessible, explaining, "A multidimensional phase space is not very easy for the simple-minded biologist to imagine or think about" (Waddington 1957, loc. 566).

We will come back to the multidimensional image later. For now, I want to suggest another way to understand his shift from the landscape to the ball. This understanding is inspired by a central epigenetic concept: canalization, the process that enables a developing organism to withstand perturbations and continue to develop in a certain way.[5] That is, an organism is robust to the shocks that might derail it from its determined direction. As Waddington theorized it, this was the evolutionary principle central to epigenetics that explained why, over the course of generations, a population tended to remain consistent. This same principle could also explain evolutionary changes: if the environmental perturbations were sufficiently great, they could push a population out of the canalized course of its development into a wholly new trajectory. In some instances, an abrupt and large-scale deviation from the canalized path of development could even be necessary for evolutionary survival. I draw on this concept to illuminate the social and disciplinary factors that shaped the changes Waddington made in his model of the epigenetic landscape.

We might think of these changes as embodying the process by which Waddington's interests were drastically redirected in response to the profound shock of the Blitz and World War II away from affectively laden images toward more comfortably neutral ones. By choosing initially to represent development as a landscape reminiscent of Britain's mountains and coastline, Waddington and Piper (whether knowingly or not) both seem to have invoked highly charged, even painful, associations: war, patriotism, and even the question of survival for the British nation. Perhaps, in discarding this realistically drawn, familiar British landscape and choosing instead to use a schematic image to model epigenetics, Waddington had indeed opted for something "easier . . . to imagine or think about."

By replacing Piper's drawing of the riverine landscape with a highly abstracted ball on a hill, Waddington clearly had freed himself from the commitment to figurative painting. He may also have been freeing himself from the historical, political, social, and geographic context of the war years. In this oscillation between situated materiality and schematic abstraction, or holism and reductionism, we can find a tension that would reappear some years later in a very different context: the elegant venue of the Villa Serbelloni, now the Rockefeller Foundation Bellagio Center, on the shores of Lake Como, Italy.

SERBELLONI AND ITS OUTSIDERS

Some twenty-seven years after Waddington introduced the image of the epigenetic landscape, twenty people gathered at the Villa Serbelloni, now the Bellagio Center, on the shores of Lake Como to work together toward a more highly theorized understanding of biology. Waddington had invited them to participate in the second of four meetings sponsored by the International Union of Biological Sciences between 1966 and 1969; their goal was to develop, "at least in broad outline, the structure of a discipline of General Theoretical Biology" that would have an "intellectual and academic stature comparable to Theoretical Physics." While the collected papers of these four meetings, known as the Serbelloni Symposia, have been described by one biologist as "a bizarre miscellany of articles about whether there can be general theories in biology, and some specific speculations about development and evolution" (Slack 2002, 894), they are in fact such a rich record of the institutional and gendered tensions at play during this attempt to build a new biology that they deserve a much closer look.

While I do not claim to offer a full analysis here, I want to look closely at the second year of the symposium, during which Waddington explored the initial impulses and influences that led him to formulate the original epigenetic landscape, as well as those that would later induce him to abandon it as a scientific model. The story concerns a series of disciplinarily fraught and gender-inflected exchanges among three participants in particular: Waddington, the embryologist-turned-geneticist; the quantum physicist David Bohm (who would later leave the strait halls of physics to engage in increasingly wide-ranging explorations of the connections between mental and physical phenomena); and Marjorie Grene, a founding figure in the philosophy of biology who moved from a collaboration with Michael Polanyi (the originator of the idea of "tacit knowledge") to an extended application of the works of the ecological psychologists Helen and J. J. Gibson.

When the second Serbelloni Symposium began in 1967, the question before the assembled participants was whether they could frame the essentials of a future field of theoretical biology. Here, as in the inaugural year, Waddington had chosen the diverse group of participants on the principle that they were "most likely to have something interesting to say" (Waddington 1972, preface). Coming from a man whose publications spanned "paleontology, population genetics, developmental genetics, biochemical embryology and theoretical biology," this was certainly an invitation for the participants to think broadly (Slack 2002, 889). Participants at the first symposium, in 1966, had decided that unlike physics, theoretical biology "would not seek for universal and eternal laws" (Waddington 1969, preface). Instead, agreeing that living systems are particular examples of some kind of "organized complexity," they saw their task as asking, "What kind of complexity?" and "What are the principles of its organization?" (Waddington 1969, preface). At the second Serbelloni Symposium, the equally diverse group of participants included Bohm, Grene, a molecular biologist, three neuroscientists, a geneticist, seven physicists, a theoretical chemist, a chemical engineer, a systems analyst, a philosopher, and an automata theorist.[6] During this second year, a consensus began to emerge that affirmed a reductive notion of biological science. If we examine the interactions at the second Serbelloni Symposium with an eye to the constraining effects of gender and disciplinarity, we not only can glimpse a short-lived moment of resistance to those reductionist tendencies, but we also can see glimmerings of the successor sciences to come, as embodied in systems biology, an expansion of embryology and developmental biology to incorporate ecology and medicine, and developmental systems theory (Gilbert 2014; Oyama 1985, 2000).[7]

Waddington's contribution to the symposium, "The Practical Consequences of Metaphysical Beliefs on a Biologist's Work: An Autobiographical Note" (Waddington 2010), illuminated the forces propelling the scientific project at the center of the Serbelloni Symposia while also indicating some of the personal and institutional forces that may have narrowed its outcome. His presentation followed papers by Bohm ("Some Remarks on the Notion of Order" and "Further Remarks on Order") and a response by Grene ("Bohm's Metaphysics and Biology"). The common theme of the discussion was metaphysics.

Arguing for what he called "The Metaphysics of Process," Bohm claimed that the desire for order is the common fundamental principle underlying physics and biology: "Things, objects, entities, are abstractions of what is relatively constant from a process of movement and transformation." Indeed, he continued, "Order may well be the basic factor which unites mind and matter, living and non-living things, etc." The metaphysics of process is a self-regulating hierarchy and "man is a part of the vast totality of all process" (Waddington 1969, 59). Bohm concluded by integrating abstract human thought into that system through a feedback loop with epigenetic overtones: "By means of these abstractions, man is able to assimilate the world within his consciousness while he is also participating creatively in the world, to help transform it, so that it will be more suitable to his needs (which in turn are changing in this process). These two movements (assimilation and creative participation) are two inseparable sides of one 'circular' process" (60).

Bohm seems to have been unaware of the uproar his paper caused by its embrace of the metaphysics of process. As he recalled it, the participants at the Serbelloni Symposium were united by "a common realization that metaphysics is fundamental to every branch of science" (Waddington 1969, 41). However, Waddington was very aware—as, no doubt, any conference organizer would be—of the negative response in the audience to Bohm's ideas. He was also in sympathy with the process philosophy Bohm articulated, and we can hear him oscillate between the two positions in the paper he presented. Waddington's conflicted response to Bohm raises very important questions about the role of the Serbelloni Symposia in the direction taken by biology in the next several decades, so I will begin with Waddington's response before rounding back to the response to Bohm that preceded it, by Grene.

THE PRACTICAL CONSEQUENCES OF METAPHYSICAL
BELIEFS ON A BIOLOGIST'S WORK

In his presentation, Waddington embraces the controversy Bohm's talk generated, siding with metaphysics, unlike "several of the more 'hard-headed' characters at the Second Symposium [who] expressed from time to time, at cocktails or after dinner, a suspicion that metaphysical considerations . . . have ultimately no real impact on the directions in which science advances."[8] Dissenting from their view of metaphysics as merely "part of the froth churned up while the theoretical physicists flounder and thrash about trying to find a firm footing in the deep and dangerous waters of quantum theory, sub-nuclear particles, and the like," he proclaims that "a scientist's metaphysical beliefs are not mere epiphenomena, but have a definite and ascertainable influence on the work he produces." And he puts his own experience on the line in support of Bohm's affirmation of metaphysics: "I am quite sure that many of the two hundred or so experimental papers I produced have been definitely affected by consciously held metaphysical beliefs, both in the types of problems I set myself and the manner in which I tried to solve them" (Waddington 1969, 72).

The richly visual paper that follows presents Waddington's "metaphysical-experimentalist's autobiography," explaining, as he puts it, the "two (or perhaps three) notions which infiltrated into my thinking at a very early stage, without much benefit of academic dignity, and which have remained there ever since."[9]

> **The world egg.** "Things" are essentially eggs—pregnant with God-knows-what. You look at them and they appear simple enough, with a bland definite shape, rather impenetrable. You glance away for a bit and when you look back what you find is that they have turned into a fluffy yellow chick, actively running about and all set to get imprinted on you if you will give it half a chance. Unsettling, even perhaps a bit sinister. But one strand of Gnostic thought asserted that *everything* is like that.

> **The Ouroboros, the snake eating its tail.** This famous symbol, which is as well known in ancient China as in Alexandria, expressed the whole gist of feedback control almost two millennia before Norbert Wiener started "creating" about the subject at MIT and invented the term "cybernetics." (Waddington 1969, 74)

As a schoolboy, Waddington drew an image of the ouroboros; he includes it in the published proceedings of the symposium with the explanation, "Inscribed within the Ouroboros is a third subsidiary notion . . . 'the one, the all,' a phrase which implies (in a cybernetic context, be it remembered) that any one entity incorporates into itself in some sense all the other entities in the universe" (Waddington 1969, 73–74) (figure 1.3). On the left, the drawing pictures the snake swallowing its tail, a vital image of circularity calling to mind the cybernetic notion of feedback; on the right, two circles reference the ancient alchemical art of abiotic creation

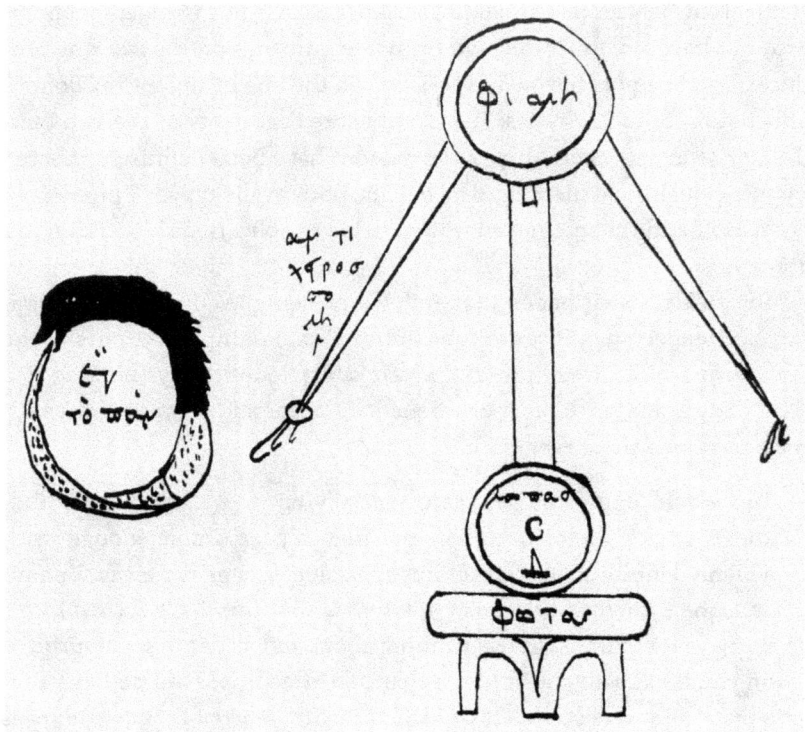

FIGURE 1.3 Waddington's schoolboy drawing, circa 1923: "The Ouroboros, together with an alembic (distillation vessel) redrawn from an alchemical document known as the Chrysopeus (i.e. gold-maker) of Cleopatra" (Waddington 1969, 74).

through which base metals are transmuted into gold. The image alludes to, but never pictures, the world egg, the dynamic source of all organic and inorganic materiality.

Consider the tensions at play in this presentation. Waddington's descriptions of the world egg and the ouroboros may have raised some eyebrows, even in this interdisciplinary audience, because of their associations with the discredited science of alchemy and the female ovum. The jumble of theatrical metaphors with which he defends the role of metaphysics in science similarly seem unlikely to convince the resistant audience he described at the outset. Metaphysics is "something more than a set of decorative flourishes on the proscenium arch, giving on to the stage in which the real action takes place," he asserts. "Metaphysics can be absorbed through communication-channels other than extended rational exposition" (Waddington 1969, 72). Waddington's metaphors, like his autobiographical disclosures, carry a gendered and disciplinary set of associations that are at odds with the shared project that has brought the scientists to Serbelloni: defining the properties of a new field of general theoretical biology.

Let me explain what I mean. Waddington believed there was a strong need for this new scientific field. As he had explained in print only a year earlier, "Theoretical physics is a well-recognized discipline, and there are departments and professorships devoted to the subject in many universities. . . . It is widely accepted that theories of the nature of the physical universe have profound consequences for problems of general philosophy. In contrast to this situation, theoretical biology can hardly be said to exist as an academic discipline" (Waddington 1968a, 525). Waddington's interest in creating a general theoretical biology was certainly shaped in part by a phenomenon we might call "P.E." I borrow the term from the biologist Lynn Margulis, who used it to designate a desire for the mathematical precision and the disciplinary dominance of physics. She recalls a time when Richard Lewontin gave a presentation on evolutionary mechanisms in an economics class at the University of Massachusetts. When she asked him why the elaborate mathematical model he presented was "devoid of chemistry and biology," she recalls, he responded enigmatically, "P.E." What could he mean, she thought. "Population explosion? Punctuated equilibrium? Physical education?" "'No,' he replied, "'P.E. is 'physics envy' . . . a syndrome in which scientists in other disciplines yearn for the mathematically explicit models of physics" (Margulis 1995, 132).

Waddington, too, was motivated by P.E., but of a different sort: philosophy empathy—an attraction to the holistic possibilities of philosophy. He reveals this in the autobiographical note he presents to the symposium (Waddington 2010). Posing the rhetorical question, "What did I actually do as a practicing biologist, and how was this influenced by this metaphysical background?" he recounts a career that seems straightforwardly focused on experimental research. The narrative moves from investigations of the ammonite as a paleontologist and experiments on *Drosophila* evolution to the final focus on experimental embryology in the series of experiments on chick embryos that led him to theorize the new field of epigenetics as a combination of the insights of embryology and genetics.

Yet rather than dwelling on this moment that made his name and embodied the metaphysical precepts embodied by the world egg and the ouroboros, he interrupts himself to acknowledge an even more important influence. "Before these highly poetic metaphysics had any practical influence on my scientific work," he explains, they were overlaid by "a large body of much more explicitly rationalized thinking": the writings of the philosopher Alfred North Whitehead (Waddington 1969, 73–74).

The distinction Waddington makes here between "highly poetic metaphysics" and "explicitly rationalized thinking" is a significant one. As an experimental embryologist who chose that field—he jokes—only because he would have been unable to earn a living in his preferred profession as a geneticist, he certainly felt the pull toward reductionism generated by the atomistic, Newtonian view of the world that dominated science at the time. Yet in Whitehead's interactionist ontology he found an exciting alternative perspective: the very raw materials of science could be understood as "occasions of experience," endless, innumerable moments of unity between "an experiencing subject and an experienced object." Whitehead's central notions of "concrescence" (i.e., "the coming together of the constituent factors in an event") and "prehension" (i.e., "the way an event here and now incorporates into itself some reference to everything else in the universe in accordance with its own 'subjective' feeling") appealed to him, perhaps because they shared the qualities of convergence and holism that drew him to the poetical images of the world egg and the ouroboros. Yet he reframes them here as "explicitly rationalized thinking" that even anticipates the mathematical rigor of quantum physics: "the idea that a particle must be thought of as a wave function extending through the whole of space time."

There is, I am suggesting, some rhetorical fancy footwork here. Although he asserts a similarity between Whitehead's philosophy and Bohm's metaphysical notion that "nature is more like an artist than an engineer," and even goes on to acknowledge that privately "my own thought runs along similar lines," he stops short of claiming them as part of his scientific practice. "I do not see that they have had any direct influence on the way in which I have conducted experimental work, which is the subject which we are discussing here," he writes. "As far as scientific practice is concerned, the lessons to be learned from Whitehead were not so much derived by his discussions of experiences, but rather from his replacement of 'things' by processes which have an individual character which depends on the 'concrescence' into a unity of very many relations with other processes" (Waddington 1969, 74–75).

How do we understand this waffling? Waddington distinguishes between Whitehead's conceptual and methodological aspects; he maintains that his own work as an experimental scientist was most shaped not by Whitehead's ideas but by his practices. What Whitehead taught him, Waddington explains, is to think not of things but of processes and to approach individuals as the coordinated convergence (the concrescence) of many such processes. It was this insight that bore fruit for Waddington in the late 1930s, in the landmark series of experiments with cultured chick tissues in which he "began developing the Whiteheadian notion that the process of becoming (say) a nerve cell should be regarded as the result of the activities of large numbers of genes, which interact together to form a unified '*concrescence*'" (Waddington 1969, 80). These very experiments inspired him to propose the new field of epigenetics. He felt that biologists needed a theory to explain the relationship between the genotype and the phenotype and hoped that unlike embryology or genetics alone, the new field of epigenetics could account for this relationship. It was to convey the broader theoretical and pragmatic capability of this new field that Waddington created the epigenetic landscape, "a visual depiction of a set of developmental choices that is faced by a cell in the embryo," expressing the way these relations shape and channel development over time (Hall 1992, 119; Slack 2002, 893).

Note the contradiction here. In his formulation of epigenetics, which is indebted to Whitehead, Waddington asserts that we should understand model experimental subjects—chick embryos, for example—not as autonomous entities but as entangled processes that are always changing in response to the environment. But in his autobiographical encounter with the world egg, he resists the understanding that "everything is like that," which he associates with gnostic thought, just as he resists Whitehead's "discussions of experiences." Approached as a simple phenomenon in cell development, the world egg arouses a complex mixture of emotions in him: mastery (he describes it as "simple enough, with a bland, definite shape"), detachment (it can be researched), and a relaxed sense of its alterity (he describes it as "rather impenetrable"). Yet understood as a reference to something real, the world egg produces surprise, even anxiety.

The fluffy yellow chick that hatches from the world egg seems "unsettling, even perhaps a bit sinister," as it actively runs about and may even imprint on him, if given "half a chance" (Waddington 1969, 73). To believe that interactions are truly reciprocal would be to apply the understanding of entangled processes to the experimenter's own experiences. This would require him to explore the impact of scientific work on the experimenter and on the human community, as well as on the experimental subject. By excluding this consideration of affect and context from his scientific work, Waddington limited the concept of epigenetics, narrowing and ultimately closing down the potential applications of epigenetic theory.

Waddington explains this retreat from metaphysics to experimental science as a strategic one. He positions himself as opposed both to applied science and to philosophy: "All of us who want to understand living systems in their more complex and richer forms are fated to look like suckers to our colleagues who are content to make a quick (scientific) buck wherever they can build up a dead-sure pay-off" (Waddington 1969, 81). Moreover, he invokes both gender stereotypes (the frightening bully) and disciplinary stereotypes (low-prestige philosophy in relation to high-prestige science) to explain this refusal to acknowledge his metaphysical sympathies: "Since I am an unaggressive character, and was living in an aggressively anti-metaphysical period, I chose not to expound publicly these philosophical views" (81).

Certainly Waddington was opting for reductionism over complexity when he backed away from acknowledging the implications of his metaphysical views. But the crucial point here is that in making that argument, he also restricts the implications of epigenetics rather than exploring those reflexive and emergent systems that harmonize with his metaphysical interests in the world egg and the ouroboros. Thus, according to one philosopher of biology, his epigenetic theory remains an expansion of "the gene-centric view in biology" that includes "a flexible and pragmatically oriented hierarchy of crucial genomic contexts that go beyond the organism" (Van Speybroeck 2002, 62). Understood in those terms, we can understand how his decision prepared the way for the disciplinarily restricted, gene-centered epigenetics of today, in which physicians screen for BRCA1 genes but have little impact on the environmental causes of breast cancer. In contrast, a broadened vision of the epigenetic landscape can generate an alternative environmental understanding of cancer etiology (Eisenstein 2001).

MARJORIE GRENE: THE OUTSIDER SEES MOST OF THE GAME

"Incredible—a scientific conference where people can talk metaphysics without being shot down!" (Waddington 1969, 338). The physicist W. M. Elsasser's comment seems at best ironic when it is considered in light of Waddington's decision to align himself with the aggressively reductionist position of the audience rather than affirming publicly the metaphysical concepts he describes as integral to his life and career. A sharper understanding of the stakes in speaking about metaphysics at the second Serbelloni Symposium appears in another response to Bohm, delivered by the only female participant in the symposium: the philosopher of science Marjorie Grene.

Grene may have been the ultimate outsider at the symposium, and as Waddington himself observed, "The outsider sees most of the game." She was as orthogonal to her discipline as Waddington was to his. "I have always been a maverick in philosophy," Grene recalled in *A Philosophical Testament*, "taken, most of the time, with unacceptable minority positions, and also (as women so often have been) totally outside professional life for many years." A zoology major at Wellesley College, Grene went on to attend Heidegger's lectures in Freiburg as an exchange student. She earned her master of arts and doctoral degrees at Radcliffe, which she describes acerbically as being "as close as females in those days got to Harvard. . . . Indeed, when I had passed my final orals for the doctorate I was told: 'Goodbye; you're a bright girl but nobody gives jobs to women in philosophy'" (Grene 1995: 5). Until she went to Manchester in 1957 to work as Polanyi's research assistant, she has written, she was involved mostly "with marriage, family, and farming." This persistently marginal relation to her chosen profession may explain her jibe, "The more eminent a person is in academic philosophy, the more likely it is that he (or she) is a fraud" (2).

Yet Grene's position on philosophy was not simply another case of P.E.—in this case, *professional envy*. Grene spent much of her career on a working farm, unlike (as she put it) "most philosophers, [who] I have found, live in a philosopher's room, where all apertures have been hermetically sealed against reality and only recent copies of a few fashionable philosophical journals are furnished to the inhabitants" (quoted in Mullins 2010, 35). Awareness of that vital world beyond the philosopher's room was a crucial value for Grene. Late in life, she described with bitterness how "research fields, disciplines with developing traditions capable, through experiment, theory, and/or observation, of adding to the body of human knowledge, have been dropping out of philosophy like flies for several centuries." She enumerated the fields lost to philosophy: the natural sciences, the social sciences, and, finally, psychology. And she mournfully tabulated the results of the losses: "What we have left, as academic philosophers, is the history of our own literatures, conceptual analysis of disciplines beyond philosophy that throw up problems we can help with, or . . . original reflection on fundamental, usually insoluble problems. The rest of our posturing is just plain rubbish, often fashionable, status-conferring, even (relatively speaking) money making rubbish but rubbish all the same" (Grene 1995, 3).

This situated knowledge as a woman, a farmer, and a philosopher suffuses Grene's response to Bohm at the second Serbelloni Symposium, despite her resistance to being described as a "pioneering professional philosopher."[10] Grene's career had sharpened in her an awareness of the forces of hierarchy and exclusion, as well as a maverick appreciation of the philosophical potential of fields beyond philosophy. Beginning with the observation, "A philosophical observer at a scientific conference is a kind of ethologist (or epistemologist?) watching the conceptual behavior of the other animals," Grene goes on to offer an assertive, acerbic, and insightful view of the scientific response to Bohm's "metaphysic of process" from the perspective of a disciplinary outsider. She issues a backhanded compliment to the participants for being "a group of extremely ingenious and well-trained performers" and then diagnoses the symposium as suffering from a clash of scientific paradigms: the one "orthodox and relatively restricted and restricting" and the other "heterodox and comprehensive." Unlike most paradigm clashes, where two subgroups simply battle each other to a stalemate, however, Grene explains that in this confrontation one group "is literally comprehended—that is, described and explained, by the second, though its members were, with some exceptions, unaware that this is what had occurred" (Grene 1969, 61).

Grene's analysis is worth quoting at length for her play with disciplinary and gender stereotypes, as well as the subtle way she describes, applies, and then performs the process she calls (with wonderful irony) "comprehension." Note the disciplinary distancing in her use of evolutionary rhetoric and the gender play in the tentative way in which she frames her commentary, invokes mock incapacity when faced with this "instructive" spectacle, and anticipates her audience's disapproval. Note, finally, the self-deprecatory apology for offering her "poor best," followed with the devastating vagueness with which she frames the scope of her commentary:

> It was—I hope—a case of evolution in action: where the species doomed to extinction, innocently unconscious of its lack of "fitness," continues happily to perform its traditional rites. The spectacle was instructive but difficult to report, for two reasons. On the one hand, David Bohm in his original paper and in his "Further Remarks" has himself indicated plainly how his "metaphysic of process" assimilates and explains the truncated metaphysics of orthodox biology (and physics and computer science and psychology, etc.). Yet on the other hand most of the contributors to and probably most of the readers of this volume, subscribing as they do to the still current orthodoxy, which as a matter of fact flourishes exceedingly at present, rather like the horns of the Irish elk, are unlikely to see the pertinence of Bohm's metaphysical remarks to their own methodology, and so are unlikelier still to see anything but *im*pertinence in my remarks on these remarks. The poor best I can offer in these circumstances is to try to put Bohm's speculations and, by implication, the metaphysics of the orthodox majority, into their historical context in terms of the major development of philosophical thought in the past three centuries or so. (Waddington 1969, 61)[11]

The attempts at biological theory building at the second Serbelloni Symposium have been incoherent, Grene charges. They lack three critical elements: "a concept of act, that is, of creative change of order"; an alternative to the "thoroughly functional" approach shared by all of the symposium; and a strategy for apprehending "emergent orders" that focuses not on "manipulation of means" but on "understanding" (Grene 1969, 61). As sharp in her implicit gender analysis as in her delineation of disciplinary dynamics, Grene attributes these omissions to "temperamental reasons: the passion for model making, and social reasons: the prestige and power granted to machine makers in our society." And she sums up their effects bitterly as a vicious circle: "To cut off mind from nature is to cut off subject from object, so sharply that science itself (the product, after all, of subjects) becomes irrational and reality meaningless. Science becomes computation-for-the-sake-of-prediction-for-the-sake-of-computation-for-the-sake-of-prediction."[12] Challenging the conventional model of biological theory, with its arid machine of computation for computation's sake, she argues for experientially based understanding. As maverick to orthodox philosophy as Waddington was to mainstream biology, Grene's outsider position enabled her to articulate an embedded, embodied, emergent vision of development that both honors and extends Waddington's concept of epigenetics (Waddington 1969, 73).

The context of conversation at Serbelloni, if only nominally interdisciplinary, was without question implicitly gendered. Although Grene singled out as fundamental Waddington's insight that "organisms *are* not simply, they *act*," Waddington performed no reciprocal gesture of respect in his commentary on Grene's presentation. Instead, after "a remark in passing to Marjorie Grene" in the form of a disciplinary dig at her professional position, he provided a longer and "more serious comment" for Bohm's argument that "function implies the trans-functional."[13] "I do not think all such 'circles' are vicious; and I doubt if we can find, and am reluctant to believe that we need, anything beyond such circular and therefore self-sufficient systems" (Waddington 1969, 69). Not surprisingly, Grene's critique of the self-sufficient, circular discussions of Serbelloni went nowhere.

Grene never went back to the Serbelloni Symposia. From Serbelloni onward, her philosophical work was flavored with a deep critique of reductionism and human exceptionalism, as well as an awareness of the major role the environment played in her life. Setting forth what she called her "ecological epistemology" in her eighties in *A Philosophical Testament*, her "apologia pro philosophia sua," she stresses the crucial role played by the place in which one lives, whatever one's ecological niche:

> To be alive is to be somewhere, responding somehow to an environment, and in turn shaping that environment by our way of coping with it. To study human practices, including language, as forms of life is to study them as activities of the particular sort of animal we find

ourselves to be. . . . To think of ourselves as alive, as part of the living world, it is not necessary to know a lot about biology. It is just necessary *not* to reduce everything there is to either machines or feelings (or to machines, feelings, or verbiage). In short, let us put what we feel, what we do, what we apprehend around us, back into the living world, where it belongs. (Grene 1995: 63–64)

An encounter that took place just two years after the second Serbelloni Symposium may explain both Grene's choice of the affect-laden word "coping" and her assertion that human beings are "the sort of animal we find ourselves to be." In 1969, Grene and Hubert Dreyfus collaborated on convening "one of those large amorphous, or polymorphic, meetings" (she wrote in 1995) "that don't, I think by now, do anyone much good— on the topic 'Concepts of Mind'" (Grene 1995, 130). Some of the same ideas were in the air in Berkeley in 1969 as had been circulating at Bellagio two years earlier, and the meeting was intended to explore them: "the philosophical aspects of work in Artificial Intelligence, and . . . chiefly in connection with philosophical questions about biology, . . . the problem of reductionism as it was then debated: is biology reducible to physics and chemistry, are minds reducible to nervous systems, and so on."[14]

At that "Concepts of Mind" meeting, Grene first encountered James J. and Eleanor J. Gibson, whose ecological psychology, she quickly felt, could "contribute to a new and more fruitful account of animals,' and particularly human animals,' knowledge of their world." From 1969 to 1995, when she published *A Philosophical Testament*, Grene developed a philosophy of biology that incorporates two concepts she drew from the Gibsons: invariants and affordances. As Grene (1995, 143) explains, three components are common to all perception: "[first,] the things and events in the organism's environment; second, the information available for pickup by the organism, usually in the form of *invariants:* constant mathematical ratios within a flowing array, and third, what the organism uses its information pickup to perceive: the *affordances* that the environment offers it." Grene's model recalls, on a larger scale, the interaction between organism and environment that Waddington explored in his theory of epigenesis and imaged in his model of the epigenetic landscape, with its attention to "mathematical ratios within a flowing array." However, Grene's model differs in one signal way: it levels the encounter among human beings, animals, and other living beings to one plane. What she has drawn from the Gibsons is the refusal of a hierarchical understanding of the relation between human life and knowledge and animal life and knowledge:

> For Gibson (or the Gibsons) as human reality is one version of animal reality, so human knowledge is one species-specific version of the ways that animals possess to find their way around their environments. Granted, our modes of orientation in our surroundings are peculiarly dependent on the artefacts of culture. . . . But culture, rather than being a mere addendum to nature, . . . is itself a part of nature. There is no culture, and therefore no human reality, and *a fortiori* no human knowledge, not dependent on the use of natural materials and itself contained in nature, in the natural environment of mother earth, on whose existence we all depend. And our efforts . . . to use language, instruments and pictorial representations to find our way in our human environments . . . are analogues of other animals' devices for finding their way in their less arbitrarily contrived environments. (Grene 1995, 144)

While Grene drew on her encounter with the Gibsons to formulate her understanding that human experience was not only shaped in relation *to* the environment but was inherently part of that environment, Waddington turned away from the relational potential embodied by the newly hatched chick that emerged from the world egg. Instead, he turned to an information-focused model of cybernetics, prompted by his interest in the ouroboros: "This famous symbol, which is as well known in ancient China as in Alexandria, expressed the whole gist of feedback control almost two millennia before Norbert Wiener started 'creating' about the subject at MIT and invented the term 'cybernetics'" (Waddington 1969, 73). The Macy Conferences were still of recent memory during the Serbelloni years, and they had spawned a debate between Wiener's focus on control and communication and a more broadly focused attention on interactive behavior that would persist into the 1970s and beyond (Clarke and Hansen 2009, 6–7). Waddington's ouroboros not only incorporated the stress of first-order cybernetics on information over materiality but also held the potential to invigorate a neo-cybernetic stress on environments and embodiment that would follow (Clarke and Hansen 2009). Waddington's responses to the ouroboros revealed his growing understanding of biological development as a troublingly interactive and recursive set of systems. They also provided another reason to evade the metaphysical implications of that vision in his Serbelloni talk.

FROM LANDSCAPE TO LANGUAGE

Even before the Serbelloni Symposia, Waddington had begun to abandon the epigenetic landscape. Although it arguably allowed for a broad and flexible representation of development, his desire to bring his embryological interests into conversation with genetics had led him, by 1962, to a model capable of more certainty. The challenge to this holistic, extensive, and visual understanding of biology enabled by the epigenetic landscape came from a narrower, linguistically based approach based on a new concept: the operon, a cluster of DNA under the control of a promoter, proposed by Jacob et al. 1960. This theory ultimately would enable researchers to predict developmental outcomes with specificity based on newly identifiable chemical triggers for gene expression and cell differentiation.

Yet if we look more closely, more than the discovery of the exemplary lac operon derailed Waddington's commitment to the visual, spatial, and fuzzily analogue model that was the epigenetic landscape. He had convened the Serbelloni Symposia in the hope of mapping the outlines of a pathbreaking new discipline of theoretical biology, but the end result was something much less far-reaching. In his conclusion to the final volume of the Serbelloni Symposium papers, Waddington does not even pretend to claim that a consensus has been achieved. Instead, he argues that the Symposium attendees were working toward the wrong end: their goal *should have been* not a "General Theory of Biology" but a more focused "Theory of General Biology" (Waddington 1972, 283). They should have been focused on two specific processes: the generation of "complexity-out-of-simplicity (self-assembly), and simplicity-out-of-complexity (self-organization)." While the catastrophe theory of René Thom, extensively described during the third Serbelloni Symposium in Thom's presentation "Topographical Models in Biology" (1969), offers the most fully explanatory approaches to complexity, Waddington swerves from endorsing that model (so profoundly influenced by his own epigenetic landscape) and sides instead with "the analogy of language." He affirms H. H. Pattee's contribution to the symposium, "Laws and Constraints, Symbols and Languages," which held that "the 'structures mediating global simplicity' which we have to search for in the theory of general biology are . . . perhaps profitably to be compared with languages: based on the primary biological disjunction between genotype and phenotype as the analogue of symbol-symbolized" (286–87).

This is a complex moment; let me move more slowly through it. Thom's model of the "catastrophe landscape" used a combination of mathematical calculation and visual imagery to explain how dynamic systems can change suddenly in response to small moments of structural instability. This model for cell development was strongly influenced by Waddington's epigenetic landscape.[15] While we can leave to others a discussion of the relationship between catastrophe theory and the various versions of chaos theory that would follow, there is one important thing to grasp about Thom's extension of the epigenetic landscape. Thom's theory came down on the side of determinism rather than randomness. More than a decade later, that position was questioned by scientists applying non-linear dynamics in fields as diverse as meteorology and medicine (Oestreicher 2007). But in an essay titled "Halte au hazard, silence au bruit" (Stop Chance, Silence the Noise [Thom 1980]), Thom resisted critiques of his determinism with discipline-enforcing rhetoric that recalls the rhetorical strategies of Waddington confronted, and bowed to, at Serbelloni: "I'd like to say straight away that this fascination with randomness above all bears witness to an unscientific attitude. It is also to a large degree the result of a certain mental confusion, which is forgivable in authors with a literary training, but hard to excuse in scientists experienced in the rigors of rational enquiry" (Thom 1983).[16] We know from the example of Grene that such disciplinary policing can marginalize or silence perspectives that might have been valuable to articulate and accommodate. Indeed, perhaps demonstrating the very power of peer pressure that Thom himself would go on to apply, Waddington does not side with Thom's topographical modeling as the best tool for a "Theory of General Biology" that would explore self-assembly and self-organization. Rather than finding the ideal model in either a picture plane or an image (as in the epigenetic landscape and the catastrophe landscape), Waddington turns to the model of language—and, specifically, language in the mode of information.

"Perhaps I should have foretold," Waddington quips in the epilogue, "when I wrote the introductory essay to the first volume, and stressed that biology is concerned with algorithm and programme, that this would be the company in which we should eventually find ourselves. But . . . I did not at all clearly see the direction in which we did actually move. I doubt if many others did either" (Waddington 1972, 286–87). However, while acknowledging the victory of a computer-based model of biology, whose sentences convey "programmes, not statements," Waddington still makes one last-ditch attempt to recast language as not information but activity—indeed, inter-activity. Posing the rhetorical question, "What . . . is this thing, a language, you refer to? . . . What is a language essentially?" he describes human language as something that evolves based on the continuing relationship between being and environment. In other words, language is neither soliloquy nor pure representation but inducement and interaction, attributes no longer unsettling but now worth advocating for because of their force and necessity:

> The fundamental form of "generative grammar" on this view is not: Noun Phrase—Verb Phrase. It is: *You → Do*. It must be the Second Person; if it were the First, *I* do, there is no point in saying anything; if it is the Third, *He* do, it becomes again a mere description, a statement. And that would conflict with the essential necessity for the "*Do*"—if that is omitted, no effect is produced, and natural selection can take no interest. (Waddington 1972, 287–88)

Admitting that "this is . . . the grossest heresy against the orthodoxy with which people have attempted to indoctrinate me since I was a young man," Waddington argues against what he calls "the extreme Fundamentalism of Logical Positivism" (Waddington 1972, 299). Instead, he frames evolution so broadly as to bring language back into the real of biology: "To the natural selective forces which bring about evolution—of language, as of anything else—it is precisely commands, producing effects, which do have sense, and it is statements about facts—atomic or otherwise—which are without significance or meaning" (288). One casualty of this rhetorical sleight of hand was, as we have seen, the model of the epigenetic landscape, with its evocative metaphoric imprecision. In its stead, Waddington went with the prediction and control-based "language-metalanguage analogy" because he felt it had the greatest potential to illuminate developmental processes. That would translate into the gene-centric view of molecular biology in the years to come.

Even if at the end of the four Serbelloni Symposia he was unable to resist the canalization by discipline, Waddington was able briefly to puncture the gender-segregated self-sufficiency of the second Serbelloni conference. Claiming the traditional prerogative of the conference organizer, he included two poems by another Serbelloni guest, a Mrs. Mary Reynolds, that indict the conference for its disciplinary and gender-based exclusivity and its mechanistic mind-set (Waddington 1969, preface). Beginning by mocking the self-importance of the conference-goers with a parenthetical line from Emily Dickinson ("I'm nobody. Who are you? Are you a nobody too?") Reynolds's "Conferenza di Bellagio" contrasts conversation about research findings, publications, academic departments, grants, and satirical assertions of scientific facts ("The velo- / city of light is constant because I choose / To say it is") with the quiet exchanges between two women who seem to be planning the conference ("You're right, a lake trip might even be wiser, / Give them a talk outside the sessions"). The poem concludes by returning to the Dickinson poem, no longer belittling the self-promoting scientists as frogs in an "Admiring Bog" but, instead, emphasizing the speaker's sense of exclusion, shared with her female companion: "Then there's a pair of us. Don't tell. / They'd banish us, you know" (318). A similar resistance to the reigning frame of mind characterizes Reynolds's second poem, "Sestina." Subtitled "Conference on Highly Theoretical Biology and Machine Intelligence," this poem juxtaposes the ancient beauty of the landscape to the debased present as a conference participant discusses the potential of a moon landing: "Directed by equations based on stars, / And disregardful both of night and day? / Was it for this he dreamed himself a bird?" (319).

By printing Reynolds's poems at the end of the second Serbelloni Symposium, Waddington repeated the critique of the blinkered adherence to scientific paradigms that he had earlier heard, but without affirming, in Grene's presentation. This final act of what might be called ventriloquism—that is, "the art or practice of speaking . . . in such a manner that the voice does not appear to come from the speaker but from another source, as from a wooden dummy"[17]—signals his decision to outsource his own undisciplined passions to more gender-appropriate mouthpieces. The field that emerged to claim the name "theoretical biology" would have none of the metaphysical and symbolic resources Waddington invoked in his "Autobiographical Note" at the second Serbelloni Symposium. Nor would it have the awareness of the gendered and disciplined constraints on knowledge articulated in Grene's talk and in the poems by Reynolds. Instead, in its reliance on "classical mathematical-analytical models, often explicitly inspired by theoretical physics," statistically based models, and "intensive computer modeling," the theoretical biology envisioned at the conclusion of the Serbelloni Symposia provided a narrow frame indeed for the genomic and postgenomic eras (Pigliucci 2012).

CHAPTER 2

A New Landscape of Thought

Behind Appearance

I have described the Serbelloni Symposia as a story of canalization, a process during which the deep channel of disciplinary training restrained Waddington's metaphysical impulses, redirecting him toward orthodoxy.[1] In contrast, another project with which Waddington was occupied during the same era reveals the intellectual and emotional expansion available in an environment without those gendered and disciplinary constraints. I turn now to his book on modern painting and modern science, *Behind Appearance: A Study of the Relations between Painting and the Natural Sciences in This Century* (1970 [1969]). As with the Serbelloni papers, this work, drawn from the Gregynog Lectures he had presented in 1964 at the University of Wales, revealed Waddington's penchant for crossing disciplinary boundaries. It, too, evoked resistance from critics. The reviews were lukewarm, from Martin Kemp's (1996, 29) "suggestive juxtaposition of modernist works with visual images in twentieth-century science," which still lacks "a more substantial historical foundation than he was able to provide" and Wikipedia's "has wonderful pictures but is still worth reading," to Brian K. Hall's (1992, 114) "enormously ambitious analysis of the relations between painting and the natural sciences in the twentieth century." Most damning of all was a review in the *British Journal for the Philosophy of Science* that attacked the book for attempting to apply Thomas Kuhn's paradigm theory simultaneously to art and science: "What is not mentioned at all is that Kuhn's notion of paradigms has been seriously challenged within its own field and to cite it simply as authority is less than responsible to those readers (especially artistically inclined readers) who know too little about the history of science to form their own judgments" (Jones 1971, 186).

Disciplinary gatekeeping aside, however, this massive, three-part, lavishly illustrated oversize volume characteristic of the 1960s-era efflorescence of coffee-table art books offers a perspective that keeps alive the affective and visual insights that were foreclosed at Serbelloni. Waddington makes no promise to offer "new theories about the nature of aesthetic experience" or to provide justifications for any hierarchy of painterly prominence (Waddington 1970 [1969], x). Rather, he argues that the new vision of the world introduced by quantum physics and illuminated by the writings of Alfred North Whitehead has shaped both scientific knowledge and human experience, leading to a radically different mode of understanding and representing individuals and society (ix–x). Focusing on the counterintuitive findings of quantum physics and the "retreat from likeness" characteristic of modern painting, Waddington argues that painting and modern science have something in common: they are both "revolts against old-fashioned common sense," reflecting "changes in the world view that have been occurring in the last fifty years [and that] are amongst the most far-reaching in the whole history of human thought" (1). Unlike the classical-analytical, statistical, or computerized visual models Massimo Pigliucci has identified with theoretical biology, the set of visual images in painting and scientific practice that Waddington discusses in *Behind Appearance* are heuristic and performative and thus reveal a set of possibilities for visual representation as a model for development broader than simply representational or deterministic uses.[2]

To clarify its role in Waddington's developing vision, we can ask several questions about this endeavor, particularly in relation to the simultaneous Serbelloni effort to "formulate some skeletons of concepts around which Theoretical Biology can grow" (Waddington 1968b, preface). What is the model of science that Waddington juxtaposes to modern painting in his ambitious volume? What do we make of the fact that Waddington focuses on the relation between science and painting in *Behind Appearance*, when his epilogue to the final volume of Serbelloni Symposium proceedings instead affirms the power of language to explain, by analogy and as an aspect of the new potency of algorithms and computer code, "the general underlying nature of living systems" (Waddington 1972, 283)? What does it mean for our interpretation of this volume that the structural principles of *Behind Appearance* are figures from ancient mythology, and how does this perspective inflect the vision of the world the volume locates within scientific and artistic practice? And finally, how stable is Waddington's assertion that scientific paradigm shifts drive cultural transformations rather than vice versa, considered in light of the comparison of these works?

What is the model of science in *Behind Appearance*? Waddington sets out his model of science in the wide-ranging first chapter, "The Image of Our Surroundings." He offers a chronology of advances in scientific knowledge proceeding from the *First Science* (Euclidean and Pythagorean science), to the *Second Science* (Renaissance experimental science), and, finally, to M. C. Goodall's *Third Science* of the present day, the worldview he finds represented by both the painters and the scientists under consideration in his study (Goodall 1965; Waddington 1970 [1969], 1–8). His description of the characteristics of the Third Science might well have been moved over, whole cloth, from the scientific doctrines shared at the Serbelloni Symposia: "the conception of biological evolution by natural selection," the "esoteric notions of quantum mechanics and relativity," and, finally, "the study of such general properties as information or organization . . . not entities in any usual sense but . . . characteristic of systems" (Waddington 1968a, 2). Yet in contrast to the tensions between classical and theoretical physicists and between metaphysics and science that Waddington identified at Serbelloni, *Behind Appearance* emphasizes that both scientists and artists have come to acknowledge the extent to which they are implicated—culturally, epistemologically, positionally, perspectivally, and ontologically—in their research. For scientists, he explains, this has meant drastically revising the notion that "science is completely 'objective,'" while for artists it has meant deepening their involvement in their work to the point that they think of themselves not as "delineating a scene, but [as] performing an activity in co-operation with their media and tools" (5).

Having leveled both scientist and artist to reciprocal engagement and co-creation with the material objects of their respective practices, Waddington introduces, in a sidebar in the chapter "The Scientists," a remarkable verbal rendition of the epigenetic landscape as a metaphor for scientific work:

Man in the world is like a caterpillar weaving its cocoon. The cocoon is made of threads extruded by the caterpillar itself, and is woven to a shape in which the caterpillar fits comfortably. But it also has to be fitted to the thorny twigs—the external world—which supports it. A puppy going to sleep on a stony beach—a "joggle-fit," the puppy wriggles some stones out of the way, and curves himself in between those too heavy to shift—that is the operational method of the scientists as he tries, with his blunt instruments—intellectual and experimental—to come to grips with the stubborn and unexpected world. (Waddington 1968a, 99)

Taking his reader "behind appearance" in the context of developmental biology, this alternative image of the epigenetic landscape reminds us that the very landscape of development, scaled from the size of a caterpillar to that of the entire biotic realm, is one complex intertwined system, as is the human web of meanings spun from art and culture to science and technology. The solid pegs and taut guy wires of the image become, in this verbal picture, not genes pulling development one way or the other but the thorny twigs of an external environment that both support and constrain it.

We will come back to this image in a later chapter, where it takes on explanatory and epistemological power in a very different context. But for now, note that in this verbal reformulation of the epigenetic landscape we can watch Waddington framing what he sees as the central scientific issue of the day—the debate over whether proper science should concern itself with measurable facts or struggle to achieve intuitive understandings—in terms of whether words or images are the foundation of scientific thought. In contrast to James Watson's and Francis Crick's approach to the structure of the DNA molecule ("seeing a word that will fit into a crossword puzzle"), he offers Albert Einstein's description of thinking: "The words of the language, as they are written or spoken, do not seem to play any rôle in my mechanism of thought. The psychical entities which seem to serve as elements in thought are certain signs and more or less clear images which can be 'voluntarily' reproduced and combined" (Waddington 1968a, 104).

Dispatching Wittgenstein's strategy of exploring facts through the medium of language, which he pronounces "not very rewarding territory

to anybody, and . . . far from the two interests with which we are concerned," Waddington instead explores scientific facts through a survey
of the images produced by the Goodall's Third Science" (Waddington 1970
[1969], 113). From "electron density contours for part of the myoglobin
crystal" and "weather map of the Northern Hemisphere calculated by
computer from theoretical equations supplied to it" to "computer simulation of turbulent flow" and "two spectrographs of bird song," he selects
images that picture biotic and abiotic systems that scale from the minute
to the vast (121–23). These examples, he argues, serve as "a new 'landscape
of thought,' a new climate of form" (119). Not only do artists now pay attention to such scientific images, they do so because they have realized
that "hard scientific analysis—from which conclusions could be drawn . . .
led to images of the same general kind" (119). Instead of a unified theory,
whether of art or of painting, all humans (whether scientists or artists)
share a process that sounds very much like Whitehead's concrescence and
prehension: the search for "a unit of thought, feeling, and action, to which
he can attach himself" (6).

**Why has he focused on the relations between science and painting rather than
science and language?** As he explains in "The Image of Our Surroundings,"
while both modes of human endeavor have the same goal—to produce
syntheses of broad scope—scientific language is so technical and specific
that it risks putting writers at a disadvantage. "One cannot discuss in
words the implication of physical theories of indeterminacy, or of biological theories of genetic determination, without the question being asked
whether the writer can understand in words what those theories state"
(Waddington 1970 [1969], 4). The point here is not simply the inability
of a novelist, wielding a nonscientific vocabulary, to grasp the technical
vocabulary of science. Rather, the point is the inherent impossibility of
expressing in *language* what scientific research reveals. In fact, as Waddington goes on to argue, comprehensive representation, whether in
language or image, is not the ultimate goal. Just as science requires "the
exercise of the faculties of insightful perception of natural phenomena
and of the imaginative creation of new concepts"—neither of which, he
argues, is an inherently verbal process—so, in his view, is painting closer
to performance than to pronouncement. If painting does comment "on
the world," it does so "not by logical or even visual analysis of it, but by a

process of 'showing' similar to that which Wittgenstein claimed was the only way of expositing to view the most profound truths of philosophy" (4). Note how different this position is to his comment in the epilogue to the final Serbelloni Symposium, which falls in with the vogue for information theory so prevalent during the period of first-stage cybernetics: "To a biologist . . . a language is a set of symbols, organized by some sort of generative grammar, which makes possible the conveyance of (more or less) precise commands for action to produce effects on the surroundings of the emitting and the recipient entities" (Waddington 1972, 288).[3]

What might be the shaping effect of mythology on the vision of the world in *Behind Appearance*? Mythic titles designate the three sections that constitute the bulk of the volume: "Part One: The Binocular Cyclops," "Part Two: The Hybrid Argus," and "Part Three: Sucklings of Diana of the Ephesians." Waddington begins by indicting the monstrously one-sided visions of the universe provided by science and painting from the time of Cubism onward. Almost as a mark of his openness to the first version of the epigenetic landscape (and all it incorporates in the way of affect and environment), he acknowledges his own contrasting preference not for abstraction but for "the rougher stuff of Piper . . . [whose] work of just after 1940 was often dismissed by the then-reigning Art Establishment as representational, and therefore retrograde in the eyes of purists for abstraction; or denigrated as 'theatrical'" (Waddington 1970 [1969], 58). The chapter reflects his opening claim that both science and painting are ideally hybrid forms, made stronger by coming together, neither needing to be restricted to one perspective on the world. Science, like painting, is "no sort of a Cyclops—monocular or binocular—it is more of an Argus, with a hundred eyes" (6). But there is a deeper role to the mythology in Waddington's view of science in this volume, which we can see even in its introductory chapter, "The Image of Our Surroundings." It expresses a still dormant aspect of his understanding of development as involving a powerful maternal presence that shapes and nurtures all living beings. In hindsight, the Gaian undertones here are unmistakable.

We can sense this presence as the first chapter rounds to its conclusion. The rhetoric recalls and exceeds the idiosyncratic vision of Waddington's autobiographical Serbelloni presentation as it participates in the contemporary fascination with notions introduced by Melanie Klein, D.

W. Winnicott, and other object-relations theorists. It is interesting indeed to speculate on whether Winnicott, innovator of the "squiggle technique" of free association, might lie behind not only Waddington's interest in sketching and painting but also this deeply primary process-laced vision of the world as a "good enough" nurturing mother:

> Psycho-analysts argue that the basic impulse for such strivings is the desire to find our way back to the good breast from which we were perforce weaned. Perhaps the main conclusion this book will come to is that, at least in the fields of science and painting, there is no one good breast to be discovered or rediscovered. The world has too much to offer for us to take in what Gide called our terrestrial nourishments from any single source. We should be worshippers of the many-breasted Diana of the Ephesians—Diana Polymastigos. (Waddington 1970 [1969], 6)

With this reference to the ancient image of nature, the Mother of All Being, we can see Waddington again oscillate away from the masculinist gender dynamics of the Serbelloni Symposia.[4] Waddington's foreword to *Behind Appearance* suggests what may lie behind that different perspective, because in it he expresses his greatest debt to one person who was particularly important to his thinking: the biologist Ruth Sager, whose "genuineness of response, coupled with a challenging capacity to expose the weaknesses of any argument or line of talk which don't quite make the grade, did more than anything else to force me to find time to put this all down on paper" (Waddington 1970 [1969], foreword).

RUTH SAGER AND THE TURN TO A SYSTEMS-BASED APPROACH

A plant geneticist who "almost single-handedly developed [the] subject of non-Mendelian, cytoplasmic genetics," Ruth Sager was present at the first and third Serbelloni Symposia but did not contribute to either volume of the collected papers (Pardee 2001, 3). In this way, and in several others, Sager's story not only recalls that of Marjorie Grene but will also be a familiar one to feminist science studies scholars. A brilliant student who earned her masters of science in plant physiology, Sager spent World War II working as a secretary and an apple farmer. Although she was highly qualified, there was no space for her in academic science for several decades. Instead, she waged her own battle from outside the field to convince her peers that the cytoplasm played an important part in heredity and that cancer came about due to the "inactivations of tumor suppressor genes" (8). Although she stood "at the pinnacle of research on the problem of non-nuclear or cytoplasmic genetics for many years" (3), it took her twenty years to obtain a faculty position. Her work on cytoplasmic inheritance and chloroplast genetics would earn her election to the National Academy of Sciences in 1977 and, in 1988, the academy's Gilbert Morgan Smith Medal (Schmitt 2008). Her research would eventually lend support to Lynn Margulis's theory of the endosymbiotic origin of organelles: "If an organelle originated as a free-living cell, it is possible that naturally occurring counterparts can still be found among extant organisms" (Schaechter 2012, 14).

Waddington's tribute to Sager reveals not only that he takes seriously the concepts she forwarded (the crucial role of the cytoplasm and the role of epigenetic factors in reproduction and carcinogenesis), but also that the environment in which he is at work shapes his capacity to acknowledge that influence. Let me be clear: I am suggesting that when he was writing about art, as well as science, and into a community of friends that included not only some of the preeminent artists of his era but also a group of very influential women that included his wife, the architect Justin Blanco-White, and her friend Myfanwy Piper, Waddington may have found it easier not only to acknowledge his female colleagues but also to remain open to concepts and methods they put forth that were less accessible to him at Serbelloni. His homage also reflects a willingness to explore the kind of systems-based approach to growth Sager's work embodied. We can see this approach developed in "The Profits of Plurality," in "Part Three: Sucklings of Diana of the Ephesians." In addition to the distinctively 1960s-era politics of this section, which advocates a shift from a philosophically unified system of hierarchical culture to "an Egalitarian Democracy of ideas and activities," the most notable aspect of this chapter is the model of scientific thinking it adapts, one that accommodates precisely the reflexivity he seems to have shied away from at Serbelloni (Waddington 1970 [1969], 238). Waddington argues that with the distinction between observer and observed breaking down, scientists of this Third Science understand how they are part of—indeed, implicated in—their experiments in ways that were never envisioned during the height of Second Science experimentalism.

Rather than being unsettled by the activity and potentially symbiotic engagements of the world egg hatchling, *Behind Appearance* describes this vital revelation, with its gnostic overtones, as a welcome aspect of the new pluralistic worldview. This insight from biology draws together the arts, the sciences, and other human endeavors in "some type of 'organicist' view, which emphasizes the unity and reciprocal interaction between man and nature, and between the various aspects of human existence" (Waddington 1970 [1969], 242). The vision fulfills the point he makes somewhat whimsically in the preface to *Behind Appearance*: "Our picture of human nature must be in quite other dimensions if we consider that the basic structure even of the physical world is such that *everything is really everywhere, though in some places more than others*" (x).

Waddington argues in the concluding chapter of *Behind Appearance* that language offers an insufficient entry point to the understanding of the universe shared by modern painting and modern physics. We must move beyond conceptual categories, including the literature-science divide, he argues, to realize the "essential artificiality" of disciplinary fragmentation. He reminds us approvingly of Aldous Huxley's critique of the narrowing effects of "book-learning" in contrast to performance. For Huxley, he points out, "training in the sciences is largely on the symbolic [Huxley means verbal, Waddington interjects] level; training in the liberal arts is wholly and all the time on that level. When courses in the humanities are used as the only antidote to too much science and technology, excessive specialization in one kind of symbolic education is being tempered by excessive specialization in another kind of symbolic education" (Waddington 1970 [1969], 241). Just as no scientist can fully grasp "the full load of meaning" contained in a work of Jackson Pollack while looking at the work "through scientific eyes," the reverse is also true: "We have been led, by a consideration of one apparent discontinuity in human experience, that between painting and natural science, to recognize that there is continuity between them after all, and that this continuity extends out into wider fields" (243). This vision, a product of Waddington's wandering across disciplines and engaging in the commentary enabled by such a crossover, anticipates the *tiers-instruit* of Michel Serres, as well as the transversal method of new materialism (Dolphijn and van der Tuin 2012; Serres 1997).

Finally, how stable is Waddington's assertion that scientific paradigm shifts drive cultural transformations rather than vice versa, then? Far from stable. In their metaphysical forms as the world egg and the ouro-boros, the nucleated cell and the cybernetic principle linking "the one" to "the all" were central to Waddington's biological research. Yet Wadding-ton's Serbelloni Symposium contributions to the project of theoretical biology understandably reflected the deterministic and eukaryotic focus of mid- to late twentieth-century biology. In contrast, inspired by Sager's work on cytoplasmic inheritance and chloroplast genetics, Waddington reached beyond that focus in *Behind Appearance* to explore what he called "a new 'landscape of thought, a new climate of form.'" He argued that what biology studied as systemic behavior should really be called "epige-netic" rather than "developmental" (a word frequently used by physicists), adding, "There is no reason in principle why epigenetic processes should not be completely deterministic, or why chance should play any role in them. In fact, it is doubtful if it does play any important part in most bio-logical epigenetic systems, such as those presented by developing eggs" (Waddington 1970 [1969], 107, 119).[5] While residual attraction to the predictable certainties of a Newtonian worldview, as well as his urgent present conviction that the epigenetic process of canalization constrains development in most cases, are certainly evident in this statement, we can also note the interesting pair of qualifications: "there is no reason in principle" and "it is doubtful." These caveats suggest to me that even here Waddington gave himself an opening—even if a very small one—to think otherwise: to imagine a concept of epigenetics that could take on systems profoundly influenced by randomness and chance. As the critique of ex-cessive specialization in the humanities and the sciences with which he concluded *Behind Appearance* indicates, a post-disciplinary perspective would drive Waddington's work in the last years of his life.

LIMINAL LANDSCAPES

Waddington's final volumes, *Evolution and Consciousness: Human Systems in Transition* (Jantsch and Waddington 1976) and *Tools for Thought: How to Understand and Apply the Latest Scientific Techniques for Problem Solving* (Waddington 1977) reveal his expansion of the epigenetic landscape to serve as a model for social behavior, psychology, ecology, landscape and urban design, and operations research and, finally, for explaining complex systems. The scale ranges from the molecular to the molar, from the indi-vidual human being to the environment, the biosphere, and the universe.

The reach of these volumes is remarkable, as if his embrace of continuity among disciplines at the end of *Behind Appearance* liberated in him the willingness to take even greater risks in his thinking.

The time was certainly right for such ventures. *Evolution and Consciousness*, which he co-edited with the astrophysicist and forecaster Erich Jantsch, reveals early glimpses of a "systems counterculture" that, as Bruce Clarke points out, would soon motivate people as diverse as Heinz von Foerster and Margulis to "move beyond mainstream doctrines and institutions . . . to detoxify the notion of 'system' . . . and to redeploy it in the pursuit of holistic ideals and ecological values" (Clarke 2012, 197; see also Clarke 2015). To *Evolution and Consciousness* Waddington contributed an introduction that considered the connection between human evolution and the evolutionary processes that predated and would succeed it. There he argued that "biological evolution, even at the subhuman level, is a matter of interlocking series of open-ended, cybernetic, or circular processes" (Jantsch and Waddington 1976, 15). The volume also included a chapter by Ilya Prigogine on the evolution of human consciousness and human social systems and one by the ecologist C. S. Holling that addressed the phase-space landscape as a way to model systems behavior. This was the sort of multidimensional system Waddington had once entertained for his epigenetic landscape, only to conclude that it was too difficult for "the simple-minded biologist" (like himself) to comprehend.[6] Hollings maintained that Waddington's concepts were useful when dealing with ecological systems. Applying the concept of homeorhesis to the human social world, Hollings concluded that for human beings to develop a workable new society, "We must learn to *live with disturbance, live with variability, and live with uncertainties*. Those are the ingredients for persistence" (91).

Despite its innovative and interdisciplinary contributors—or, more likely, because of them—Waddington's final contribution to *Evolution and Consciousness* has a retrospective quality. In "Concluding Remarks," he returns to the Serbelloni Symposium, as if he is trying to rewrite its outcome in this new context with this new community of interlocutors. Recalling "a series of discussions on theoretical biology . . . organized several years ago at the Villa Serbelloni in Italy," Waddington returns to Howard Pattee's notion "that we regard certain biological molecules as messages; that is to say, we consider them as conveying instructions of a kind comparable to the instructions which can be given in a natural language." While acknowledging that "this is an extremely important manner of regarding living things and their evolution," he describes himself as even less willing now than he was at Serbelloni to affirm it: "I confess I am not at all clear about the relations between sets of instruction in general, those sets which may lead to deviation-amplifying systems, and those which give rise to the self-transcendence characteristic of natural languages" (Jantsch and Waddington 1976, 248). Since information cannot answer that question, which is really one for "the professional philosophers to decide," he implies it may not provide an adequate model for biology. Thus he joins his fellow contributors to *Evolution and Consciousness* in calling for "a conscious, devoted, but critical attempt to create new guiding images of man and the future" (248).

Recanting the language-metalanguage model that would soon slide so easily into the code-based view in which genes create difference (Keller 2015), Waddington instead returns to his identity as an embryologist and selects a model based on that experience. Describing himself as an embryologist with "extensive experimental and theoretical acquaintance [with] the analysis of embryonic development," he concludes that "embryos, like ecosystems, are multifactorial" (Jantsch and Waddington 1976, 243–44). The appropriate model for biology should incorporate the tensions and constraints on development that are embodied by both an embryo and an ecosystem. Thus, he advocates the adoption and integration of multiple viewpoints, or what Jantsch calls the "symbioticization of heterogeneity" (249).

Waddington's final two books, *Tools for Thought* (1977) and *The Man-Made Future* (1978), published after his death and with a foreword by Margaret Mead, both address what we now call wicked problems: problems that are too complex and many scaled to be successfully addressed using one disciplinary approach. Waddington refers to them as the "Total Problem" and then the "World Problematique." The foundation for both books is an urgent conviction that "the ways of looking at things that we have in the past accepted as common sense really do not work under all circumstances, and . . . do not match the type of processes which are going on in the world at large" (Waddington 1977, 11).

Addressing "The Epigenetic Landscape of Human Society," *Tools for Thought* situates the topic in a discussion that moves from philosophy

(natural and moral) to the analysis of complex shapes, systems, and processes, feedback, stabilization, analysis, communication, management,
forecasting, and, finally, systems modeling ("The World as a System"). As
he describes a nameless era in *Tools for Thought*, it anticipates the notion
of the Anthropocene but deepened by a symbiogenetic awareness that
recalls the work of Sager and anticipates that of Margulis:

> The change which has occurred, or is occurring now, is that the effects
> of human societies on their surroundings are now so powerful that it is
> no longer adequate to concentrate on the primary effects and neglect
> all secondary influences. . . . The scale of very many of the impacts of
> mankind on the world surrounding him is now so great that they go
> right below the surface of things. At the deeper level, we find that most
> aspects of life and its interactions with its surroundings are intercon-
> nected into complexes. No powerful action can be expected to have
> only one consequence, confined to thing it was primarily directed at.
> It is almost always bound to affect lots of other things as well. . . . We
> need nowadays to be able to think not just about simple processes but
> about complex systems. (Waddington 1977, 11–12)

The volume's focus on finding ways to address complex political and environmental problems has led one scholar to recommend *Tools for Thought*
to climate scientists and systems analysts, despite the fact that the book
does not explicitly address either global warming or climate change. Yet
despite its resemblance to contemporary works on the Anthropocene era,
this volume is still very much of Waddington's time (Nerlich 2015). The
feel of the book is dizzying, recalling the 1970s-era sense of a social world
on the edge of transformation. Working with his collaborator, the illus-
trator Yolanda Sonnabend, Waddington introduces visual models in rapid
sequence as he addresses the problem of stabilizing human societies so
that the catastrophes that occur in that system "are little ones, not big
ones" (figure 2.1).

A tumbling flood of images—the epigenetic landscape, a J-shaped
curve, an S-shaped curve, "the final edge between the land and sea in
a great river delta like those of the Mississippi and the Nile"—conveys
an urgency that is personal as well as sociopolitical, as if the oscillation
among various disciplines with which Waddington had been struggling
for years had finally come to his consciousness:

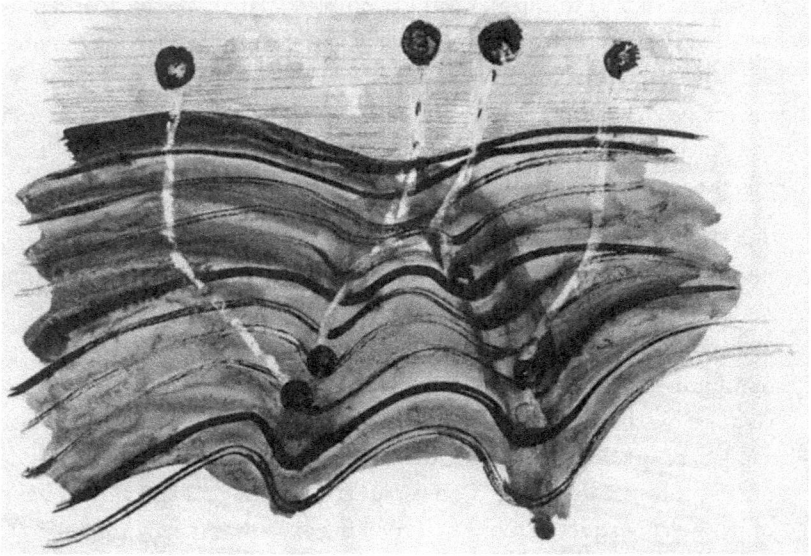

FIGURE 2.1 Yolanda Sonnabend's drawing of an epigenetic landscape that has "branching points at which a valley splits up into two or possibly more branches." Waddington comments, "Many progressive systems in fields other than biology behave in a similar way" (Waddington 1977, 109–10).

> That is to say, when we make a switch in life-style (from a junior advertising executive to a pop group/lyric writer, or from an experimental biologist to a futurologist, philosopher, or art critic), people do want the styles to be really *different*, genuinely alternative choices, not just the same mixture as before with a trifle more or a trifle less bitters in the cocktail; but surely what we want to aim at is a "system" which allows us to do this without too much danger of our whole personality being torn into shreds in the process of transition. (Jantsch and Waddington 1976, 116)

How, indeed, does one make the transition from an experimental biologist to a futurologist, philosopher, or art critic without having one's whole personality torn to shreds in the process of transition? A "system" that permits us to do this would be Waddington's epigenetic landscape writ large, but with one crucial twist: it would be open to randomness and chance, to the nonlinear vision he explored in *Behind Appear-*

ance and that he attained in *Tools for Thought*. The risks he took to get there—entertaining a variety of new ideas that "have not yet crystallized into conventional and universally accepted valuations"—were considerable, Waddington admits. "We have been discussing a field which is still very free and open-ended—free for people to make mistakes, as well as free for people to produce very fertile new insights" (Jantsch and Waddington 1976, 231).

"The limit of the frontier designates, on this side of it, familiar lands, . . . but a voyage pulls and drags this third place throughout the whole space that is thus divided. Before the frontier, less at home already than usual, the novice swims or is displaced toward the strange," writes Michel Serres (1997, 162). It is brave indeed to wander across disciplines looking for that educational *tiers-instruit*, that undisciplined third space where one can think strange thoughts and even make mistakes. Little wonder, then, that Waddington describes *Tools for Thought* in its epilogue with poetical whimsy that brings to mind *The Troubadour of Knowledge*:

> [A] book such as this . . . is a manual provided for someone who had spent his whole life in the centre of a large area of solid land . . . and who finds himself for the first time on the shores of a sea. The manual sketches out for him a large number of ways of travelling about on or in water: swimming, by breast stroke, by crawl, by sidestroke, by backstroke and so on; canoeing; rowing, sailing in single-man dinghies; snorkeling; scuba diving; and the last chapter of the book . . . might even be considered as an introduction to mechanical propulsion in large surface vessels or submarines. (Jantsch and Waddington 1976, 231–32)

THE EPIGENETIC LANDSCAPE AFTER WADDINGTON

In 2014, during a workshop on in vitro fertilization at Cambridge University, I found myself on the shores of much the same sea that Waddington faced at the end of *Tools for Thought*. I was with a group of reproductive endocrinologists, historians, social scientists, feminist theorists, and artists who had come together at the invitation of the sociologist Sarah Franklin, the historian of science Nicholas Hopwood, and the physiologist Martin Johnson to ponder what seemed initially a very simple question: "What did Robert Edwards see when he looked at an embryo?" The question seemed—no surprise here—inherently epigenetic to me, because I had been reading and writing about Waddington for some time. It made me think of his epigenetic landscape, but it also held some of the tensions facing Waddington in the 1960s, during the years of the Serbelloni Symposia and as he was writing *Behind Appearance*. Should we consider not only what Edwards saw, but also which gaze he was directing at the developing embryo before him? Were there multiple gazes? The scientific, aesthetic, historical, sociological, feminist? Were such discrete categories helpful or even possible? And what about the technologies used to make that embryo visible? How did the medium (or media) of representation shape what Edwards saw then, or what we see now?

I will come back to some of these questions in chapter 3 but want to close this chapter by returning to the paper I encountered at the workshop, passed to me not by its author (who was not present) but by a fellow attendee who thought it might intrigue me. The philosopher of science Jan Baedke's newly published, comprehensive survey of the influence of the epigenetic landscape during Waddington's own time and in the years since reminded me once again of just how profoundly disciplines constrain the gaze. In his article, Baedke traces the legacy of Waddington's model, arguing that although tradition of modeling using landscape images seemed to fall out of favor in the late 1970s, with the end of an interest in catastrophe theory, scientists in recent decades are returning to the epigenetic landscape, but with a difference. They are mining it methodologically rather than conceptually, drawing on its "double extensionality," using it as a tool to help them visualize a subject or theory, to communicate across disciplines, to stimulate creative thinking, and to devise models (Baedke 2013, 759).

In the post-Waddington era, this article reveals, Waddington's visual model has had a broad influence on life science fields that range from developmental, systems, and cell biology to evolutionary biology, developmental psychology, sociology, STS, and economics. Even before high-throughput gene-sequencing procedures produced massive amounts of "big data," Baedke reveals, scientists were returning to the epigenetic landscape not only because it enabled them to visualize complex interactions at large scales, but also because it enabled their creativity, communication across scientific disciplines, and model making. The article presents a taxonomy of the methodological legacy of the epigenetic landscape with classifications that extend across twelve life-science fields. Yet with the exception of applied mathematics; sociology; science, technology, and society; and economics, not one of those fields lies outside the life sciences, much less in the arts or humanities.

I found myself fascinated by an observation Baedke relegates to a footnote, however: "Images stimulating visual thought do not have to conform to conventional pictorial forms of expression, known, e.g., from intellectual books—a feature Waddington made extensive use of in *Tools for Thought* (1977)." Singling out Sonnabend's "rather artistic" paintings in *Tools for Thought* as intended to "stimulate the viewer's imagination," Baedke (2013, 760) then comments, "Interestingly, this fashion of unconventional illustration reappraises [*sic*] in contemporary epigenetics." I am grateful to Baedke for pointing out this phenomenon. Yet it is far more than merely interesting: it merits close examination.

What could such "unconventional" illustrations have to do with the epigenetic landscape? Like John Piper's illustration that first embodied the epigenetic landscape image, Sonnabend's paintings in *Tools for Thought* stimulate the viewer's imagination because they are visual, situated, material, and unconventional. Similarly, Eva Jablonka and Marilyn J. Lamb incorporate the whimsical illustrations by Anna Zeligowski and the delightful comic avatar, Ipcha Mistabra, in their exploration of *Evolution in Four Dimensions* because of the modes of thinking they make possible (Jablonka et al. 2014). These turns to the visual image recover and reactivate qualities contained in Waddington's original models of the epigenetic landscape, I argue, reinvigorating the potential of its fuzzy ambiguity beyond even the capacities of contemporary life scientists to exploit. In the chapters that follow, we will sample some of the uses of the epigenetic landscape in fields beyond the life sciences. As these fields incorporate "unconventional illustrations" and the modes of thinking they make possible, they mine the epigenetic landscape methodologically, using it as a tool to help them visualize a subject or theory, devise models, and communicate across disciplines. In so doing, they recuperate the full vitality of this model as a stimulus to thinkers across the disciplinary landscape.

CHAPTER 3

Embryo

Consider this image, probably the most familiar version of the epigenetic landscape (figure 3.1). When I showed it to a feminist colleague, who knew quite well that it represents a fertilized egg and that the slope on which the egg balances denotes the statistical probabilities governing its development, she remarked on how little it resembles an embryo. This image is highly abstract, even decontextualized, she said. Where is the body of the pregnant woman who plays such a crucial role in the embryo's development? She wondered how this dry visual image could represent the field in which C. H. Waddington hoped to meld the richly material findings of embryology with the promising new potential of genetics.[1]

As she spoke, I recalled a poem that Waddington wrote when he was taking a break from his experiments on avian embryos in January 1937.[2] A verse portrait of the embryo, the "bubble blastula," this whimsical lyric captures the moment a blastocyst implants in the uterine wall and begins to differentiate. Waddington gives us an embryo's-eye view of development in situ. The embryo is shown striving to develop, to "better its situation." Notice the pun on *situation*: not only is this the British term for a job, but it also suggests that a developing embryo has a specific location. An embryo is *situated*. The embryo lies between "the opportunist roof" and "the sodden floor, / with a space between." In other words, this ball of some dozen cells (comprising the trophoblast; the inner cell mass, or embryoblast; and the hollow space of the blastocyst cavity, or blastocoele) has implanted in the endometrium. In the woman's uterus, and because of a woman's uterus, a pregnancy has begun.[3]

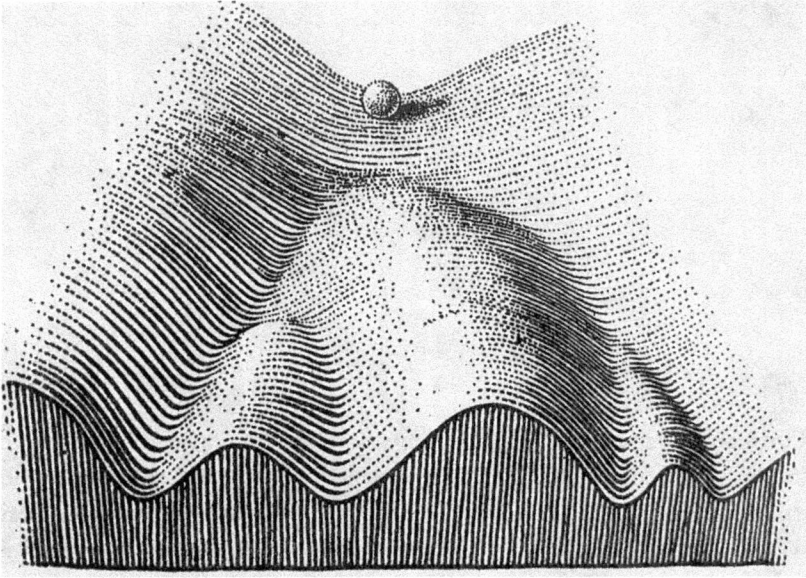

FIGURE 3.1 The second, and most familiar, version of the epigenetic landscape. Waddington's caption: "The path followed by the ball, as it rolls down towards the spectator, corresponds to the developmental history of a particular part of the egg. There is first an alternative, towards the right or the left. Along the former path, a second alternative is offered; along the path to the left, the main channel continues leftwards, but there is an alternative path which, however, can only be reached over a threshold" (Waddington 1957, loc. 591–98).

There is no visible embryo in the first version of the epigenetic landscape, the drawing that Waddington commissioned in 1940. As we know, John Piper's charcoal sketch merely figures a river, its brushy banks containing an ever changing flow. The embryo appears only in the second version of the epigenetic landscape, as a ball perched near the top of a slope whose contour lines define channels down which it can be expected to roll, given the force of gravity that is an implicit presence in this image. Just as the happy bubble blastula upstages its maternal context, those uterine walls making its ever improving situation possible, so here, when the embryo and its process of development take center stage, the enabling, surrounding landscape seems to have disappeared.[4] Instead, this second version presents a schematic image of a ball, and the slope on which it rests merely indicates a gradient in time rather than space. The very domain this image invokes seems composed of properties different from those of our everyday world. The force compelling the embryonic cell to attain its cell fate by rolling down the tilted slope is a set of probabilities embryologists must factor into their experiments, not a familiar physical property, gravity, encompassing the viewer as she orients herself to understanding how the image figures the developmental process (the roll down the slope) still to come.

This double move, eliding the mother and emphasizing the embryo, has a long history, of course. Before assuming a central position in the epigenetic landscape, the embryo was subject to the discourses, images, and technologies of the fields of observational and experimental embryology. These fields produced an embryo that is mediated, animated, and scaled, terms I will unpack in what follows.

Embryology mediated the embryo and thus laid the foundation for our sense that formal developmental regularity, discrete individuality, and implicit teleology are central to what embryos are and can be. Embryology animated the embryo to impress viewers with those foundational properties of embryonic life. Finally, and most significant for my purposes, embryology—when brought into conversation with genetics in the new field of epigenetics proposed by Waddington—scaled the embryo, framing it in multiple temporal and spatial dimensions. This final mode of working with the embryo has the potential to reclaim the cultural context, materiality, and processual engagement that my friend found so strikingly absent in the embryo she encountered in the second, iconic image of the epigenetic landscape.

MEDIATED

The visible embryo is mediated. I use this in the deliberately doubled sense, both biological and social, of a spatial or social context that makes something possible, that lets it engage in vital transformation, or that allows it to realize its potential. The term "media" dates back to Francis Bacon and Isaac Newton, the natural philosophers of the seventeenth and eighteenth centuries, for whom it conveyed "the material space that enabled the transmission of something between two points," yet by the nineteenth and twentieth centuries it had also come to mean several divergent things. Media were the means of communication by which a message could be sent and received, and they were also culture media, the fluid or solid spaces within which a living thing could grow, while the concept that something was mediated also had come to mean that it was conveyed, translated, filtered, or framed in the course of communicating it from one site to another (Mitchell 2010, 95). While all of these concepts are at play in the way the embryo is mediated, we begin with the assumption that the embryo is (and since its inception as an object of biological study has been) visible.

Moreover, the embryo's visibility is mediated through a framing or structuring technology. I use the term "technology" here in the sense that Teresa de Lauretis did in her pathbreaking study of technologies of gender to include not only objects whose technology is still apparent to us (such as microscopes, film cameras, and ultrasound machines), but also material objects or social processes where the technology is no longer evident (de Lauretis 1987). This includes modes of thought, such as philosophical or theological models of the universe; simple modes of physical representation, such as wax, clay, and metal models; and modes of conceptual representation, such as maps, diagrams, computer models, and, of course, the epigenetic landscape itself.

In what follows, I draw on Tatjana Buklijas and Nick Hopwood's excellent website "Making Visible Embryos" to show not only how embryos have been rendered visible during the history of embryology, but also how visible embryos have been made, or constructed.[5] If we visit an art museum such as Berlin's famous Gemäldegalerie, where ancient altarpieces crowd the walls, we will see embryos all around us, in sacred images such as Konrad Witz's fifteenth-century "Saint-Esprit: Der Ratschluß der Erlösung," which represents the Virgin Mary with the Christ Child clearly visible inside her. Religion mediates—that is, shapes, frames, interprets—the embryo in Witz's work so that what we see in the embryo is the artist's construction, and reflects his religious vision. Witz's Christ Child looks less like an embryo than like a pretty little boy, standing with arms outspread and gazing right out of Mary's womb at the devout viewer.[6]

Beginning with Aristotle's practice of cutting a window in the eggshell through which he could catch a glimpse (if a partial one) of the embryo as it developed, there have been seven stages of embryo imaging (Maienschein 2014). In this first stage, Aristotle imaginatively constructed a "hypothetical embryo" that accorded with his position in the preformation-epigenesis debate: that the embryo developed over time, with its form emerging gradually, rather than beginning as a fully formed, if tiny, individual (Maienschein 2014, 1). Of course, even for Aristotle, the embryo was also visible. He did not just hypothesize about it; he examined it with his own eyes, opening fertilized hen's eggs at regular intervals to study the changes he found in the embryonic chicks.

After the "hypothetical embryo" came the "observed embryo," the "experimental embryo," the "inherited embryo," the "evolved and computed embryo," the "visual human embryo," and, finally the "engineered/constructed embryo" of our era, which we can culture, study, and even build or engineer to our specifications (Maienschein 2014, 22–23). A more compact description of this timeline would emphasize three stages: classical descriptive embryology (which includes comparative embryology), experimental embryology, and developmental genetics (Gilbert 1991). Although these represent distinctly different perspectives on and ways to approach the embryo—the anatomical, the experimental, and the genetic—none of the stages is discrete. Indeed, none is *over*. They all persist to some degree in our individual and collective imaginations as we deal with embryos in scientific research, public policy, literature, and art. Moreover, as the embryo is mediated, animated, and scaled as the object of scientific study, our relation to the embryo has moved from wonder to awe, curiosity, and finally workmanlike practicality, and from detachment to instrumentalization (Gilbert and Faber 1996).

Different as they are, all of these perspectives originated in the strategy of serial representation. "Seriality" is the term for an arrangement that displays things according to similarity in an order linked to a sequence. While seriality had long been a mode of useful display in a range of fields (and I will have more to say about this later in this chapter), seriality came to scientific notice around the 1800s (Wellmann 2017). The German anatomist Samuel Thomas von Sömmerring's "Images of Human Embryos," which appeared somewhat earlier, were the first sequential images of embryological development.[7] Sömmerring was dissatisfied with the existing embryo images of his day because they did not make it possible for the viewer to imagine the seamless growth of the embryo as time passed. So working with the draftsman and modeler Christian Koeck, he created an image that presented embryos, numbered sequentially, on the same page, moving from left to right in four lines. The image was later engraved by the Klauber brothers and published in Sömmerring's *Icones embryonum humanorum*. The embryos are represented in rows, facing right, so that the flow of the gaze of reader and embryo are linked, moving with the embryo from top left to top right, and then down a row, again reading left to right, until the final image at the lower right. The first row of images shows the embryo still in utero, with the uterine tissue a cloudy aura around the internal space; images one and two even include an enlargement of the embryo, perhaps because otherwise it was too small to pick out the relevant details. All of the embryos face to the right, and genitals are clearly visible from the second row onward. Some of Sömmerring's embryos are male, and some are female. Size is clearly the principle of arrangement rather than individual developmental progress.

Sömmerring's embryo images anticipate later modes of seriality, some scholars argue, because he used them not neutrally, to represent a fact, but ideologically, to make a case or take a position in a controversy. For example, Buklijas and Hopwood argue that Sömmerring's embryo images took the side of epigenesis, or the notion that the embryo developed from an undifferentiated mass into greater and greater complexity, in the debate between that concept and preformation, the notion that the embryo exists preformed from the moment of fertilization.[8] They point out that Sömmerring noted the existence of intermediate developmental stages, claiming that those stages had their own, particular value. "Yet, he argued, do we not find a rosebud as beautiful in its own way as a rose?"[9] In their view, this image aims to express a unified and metaphysical truth about the nature of the embryo rather than include any individual variations or imperfection that a specific embryo specimen might display. Such is the rhetorical mediation of embryo imaging.

Sömmerring's image represents embryonic development according to the principle of "truth to nature," the notion that the most satisfying scientific images are the ones in which the draftsman or artist works with the scientist to standardize and idealize the image, omitting any visible flaws or imperfections. Tension between the goal of attaining "truth to nature" against later goals of "mechanical objectivity" and "trained judgment" reflected the changing position of visual images in science (Daston and Galison 2010, 18). Were images expected to make the ideal manifest? Were they required to represent what was actually there, beautiful or not, typical or atypical? Or were they supposed to select the most meaningful, while still accurate, image, selected and positioned by the expert eye? In Sömmerring's time, the connection between scientific and artistic images was relatively unproblematic: it was understood that nature itself was ordered, perfect, and thus beautiful. As he explained, "Since the anatomic description of any part, generally speaking, is just as idealistic as the representation and description of that same organ in a sketchbook, so one should follow the same principle in describing it. Everything that the dissector depicts with anatomical correctness as a normal structure [*Normalbau*] must be exceptionally beautiful" (quoted in Daston and Galison 2010, 102).

Whether Sömmerring was affirming or hesitating to affirm the principle of epigenesis, his delight in the unity, simplicity, elegance, and evanescence of the visual patterns revealed in the embryo indicate his affirmation of an aesthetic shared with many embryologists. Scott Gilbert and Marion Faber have described this as a biological typology grounded in attention to temporal change. Productive above all else of wonder, this "embryological aesthetic" was visual, emphasizing "the classical concepts of form, symmetry, pattern, integration, and harmonious interdependence," and conceptual, shaped by its scientific context (Gilbert and Faber 1996, 141). The embryological aesthetic was grounded in the classical vision of German Romanticism, delineated memorably by the biologist Paul Weiss as having three essential principles. First, living things contain all of the elements of classical beauty: "symmetry, balance, rhythm, and . . . a pleasing ratio of constancy and variety." Second, the beauty of natural forms appeals because it reveals the "measured orderliness of the developmental actions and interactions by which it has come about." Recalling Goethe's description of architecture as "frozen music," Weiss claimed, so "in the same sense, organic form is frozen development; the formal beauty reflects the developmental order." Third, Weiss argued, the orderly sequence of embryonic development expressed the beauty not of fixity but of freedom. In a significant phrase to which we will return later, he described this as "freedom of excursion," capturing the fact that embryonic tissues actually *travel* spatially as they develop (quoted in Gilbert and Faber 1996, 130).[10]

As embryology began to be taught in universities, first the magnifying lens and then the microscope made it increasingly possible to examine embryonic development in close detail. This new ability to see the embryo more clearly lent energy to the field's long-standing debate over whether the tiny embryo is furnished with all of its organs from the very beginning (preformation) or develops gradually out of nothing (epigenesis). While some scholars believe that Sömmerring aligned himself with the epigenetic position, epigenesis gained even more adherents when the development of better microscopes and new dyes to stain the delicate embryonic tissues enabled three German embryologists, Christian Pander, Karl Ernst von Baer, and Heinrich Rathke, to document the formation of different embryonic tissues. Their work revealed the origin of embryonic regions in the three different germ layers: the ectoderm, which produces the skin and brain; the endoderm, associated with the digestive tube and lungs; and the mesoderm, from which spring the blood, muscles, bones, kidneys, gonads, and heart (Gilbert 2014, 15).

While the visualization technologies of the nineteenth century helped to shape the embryo for students who had to learn microscopy to see it, so, too, did the Romantic idea of the cosmic unity of nature, for embryo mediation was more than merely technological and scientific. This is expressed vividly in the "great egg of nature" by Georg August Goldfuss. A schematic egg, shown small side up with three stacked horizontal sections in *Ueber die Entwicklungsstufen des Thieres: Omne vivum ex ovo* (1817), this image reminiscent of Waddington's World Egg "stood not just for the beginning of each living being, but also for the unity of animals on our round—if not quite egg-shaped—planet."[11] Goldfuss "drew on the Romantic concepts of analogy, polarity, and hierarchy, to place six main animal groups—Protozoa, Radiata, Mollusca, Pisces, Mammalia and Homo—onto an 'animal globe' flattened at the poles into a semblance and symbol of 'an Easter egg.'"[12] Goldfuss exhibits a remarkable attempt to merge geographical, biological, and theological meaning systems with the Romantic embryological aesthetic, exemplifying Ludwik Fleck's insight that old thought styles persist into new thought communities.

Embryological learning occurred not only when visual images and information passed from professor to student, but also when popular lecturers, entertainers, and museums presented embryological information to the general public. We turn now to a case that exemplifies this additional mode of embryo mediation and ushered in the evolutionary era of embryo investigation: that of Ernst Haeckel (1834–1919). Haeckel was a follower of Darwin known for his theory that "ontogeny recapitulates phylogeny," or that the development of an individual being replicates in its stages the evolutionary development of a species. Haeckel's images grouped embryos of different species together in a roughly chronological order to illustrate that notion, called the recapitulation principle. Many scholars have contributed valuable analyses of the controversy that occurred when repetitions of several of the plates led to Haeckel's being accused of fraud (Hopwood 2015). Yet what is most interesting to me about the Haeckel case is the bright light it sheds on the changing demands on, and uses of, scientific images when they circulate beyond the laboratory into the public realm.[13]

Another embryo debate, this time between Haeckel and a contemporary, the anatomist Wilhelm His, reveals the epistemological impact of other ways to engage with embryos beyond rendering them visible, such as mechanical modeling.[14] Drawing on his experience of making three-dimensional models of embryos, His advanced a mechanical argument for embryological development that stood in sharp contrast to the phenotypic, evolutionary argument that preceded it. In the years between 1860 and 1880, this new attention to the mechanics of embryo development was one of the forces that shifted the field from descriptive to experimental embryology. While His began with visualization, which led to insights, only to be followed by more accurate visualizations, the experimental embryologists who followed him would start with theories that they would proceed to test experimentally (Hopwood 1999, 494–95).

His's work marks a moment in which attention to multiple aspects of an embryo increases the embryologist's capacity to incorporate, understand, and remember his topic more fully and thus to communicate that knowledge more easily to a lay public. (Notably, at that point both kinds of communication were still assumed to be part of the scientist's job.) Contrasting with the detached observation of other embryologists, His's contributions were to develop the microtome, a technical tool that enabled a scientist to cut extremely thin slices of the embryo that could be studied, sequenced, and recombined into a three-dimensional model. This made it possible for an experimenter to engage actively with the embryo in multiple dimensions (or from multiple aspects). His felt that this mode of working with embryos was better because it was a richer way to produce embryological knowledge. Hopwood (1999, 482–83) says that His both argued for and performed a "doubly embodied knowledge"—the kind of knowledge produced when scientists "work not only with their brain and eyes but also with their hands. . . . Working on the same object *from different aspects* over a period of weeks built up a three-dimensional mental image. There was no substitute for this experience. . . . The modeler His agreed with the artist Ecker that 'the pictures in the memory that have once made their way through the hand stick much more firmly in the head.'"

The multiple perspectives and transdisciplinary communication His advocated were gradually rendered invisible, Hopwood (1999) argues, as the field of embryology developed. His's expert "freehand" modeling, challenged as more subjective, was superseded by the normal plate, a more generally accessible, objective technique of stacking carefully measured microtomed embryo sections together to build up a visual model.[15] The introduction of the normal plate led to a significant shift in embryology, as the field began to standardize its modes of embryo collection and exhibition (Hopwood 1999). As would occur later in the case of Waddington's thinking on biological development, the forces of disciplinary canalization gradually excluded the three-dimensional, mechanical, artisanal approach to descriptive embryology. This exclusion shaped our understanding of the embryo, as well as of the larger life narrative in which it played a part.

ANIMATED

In addition to being mediated by a range of technologies, the embryo has been animated over the course of embryological history. I mean several different things by the term "animated." First, the theological or philosophical embryo was understood to be animated in the sense of being ensouled, or infused with a vital essence, from whence its development proceeded. While the former concept was theological and the latter, philosophical, they both embraced the side of vitalism in the mechanism-vitalism debate. Another sort of animation was also important in this period, of course: the experience of quickening, the fetal movement that announced to a woman that she was "with child."[16] This particularly personal experience of embryo animation would cease to be medically significant as embryonic visibility became increasingly technologized and the maternal experience was occluded (Duden 1993).

With three-dimensional models came another mode of animation: serial spatial representation.[17] The direction of seriality evolved over time, as did the qualities indexed. The wax models of embryos exhibited at La Specola, the Museum of Natural History in Florence, show a simple increase in size rather than any perceivable developmental changes; they were arranged to move from the bottom right to the top right. In contrast, Sömmerring's embryo sequence, and the wax relief based on them by the Swiss modeler Josef Benedikt Kuriger, both invited the viewer to animate the embryo in the imagination, moving his or her eye across the images in a left-to-right, top-to-bottom movement that may have participated in the visual conventions of reading to make the embryological transformation come alive. When three-dimensional models were sidelined in preference for two-dimensional images like von Baer's ideal figures of chick embryo development, still another representational convention came to the fore, showing not a sequence of embryos but, rather, a series of embryo groupings. Instead of left-to-right scaled progressions, these visual images relied on representational conventions such as "arrows and dotted lines [to] represent movements and directions of development."[18] His saw the act of modeling itself as an important mode of reasoning.[19]

The visual conventions that shaped the understanding of embryos expanded in the late nineteenth and early twentieth centuries as many embryologists were becoming interested in Japanese and Chinese art. The embryologists Viktor Hamburger and Hilde Mangold studied Chinese and Japanese art as graduate students, and the embryologist Richard Goldschmidt was an enthusiastic collector of Asian sculpture who admired Ikegami temple gardens as "the most beautiful piece of landscaping imaginable" (Gilbert and Faber 1996, 134). Most remarkable of all, after publishing two histories of embryology and a study of morphogenesis called by one reviewer "Needham's Magnum Opus," the embryologist Joseph Needham devoted the rest of his life to studying Chinese civilization, writing a series of volumes on Chinese science, technology, and medicine.[20] Yūgen, a Japanese concept that incorporates the mysterious and profound— or "the depth of the world we live in, as experienced through cultivated emotion"—helped to shape the embryological aesthetic (Parkes 2011).[21] For example, the aesthetic of Yūgen colors the English translation of a footnote to Aristotle's De Generatione Animaleum by the Cambridge classicist A. L. Peck in 1949: "Does the embryo contain all its parts in little from the beginning, unfolding like a Japanese paper flower in water (preformation), or is there a true formation of new structures as it develops (epigenesis)?"

With its focus on the "cloudy impenetrability" of the embryo, Yūgen influenced modes of thinking about and working with embryos. So did two other prominent aspects of Japanese aesthetics: the notions of *Mono no aware* (transience) and of *Kire* or *kire-tsuzuki* ("cut-continuity"; Parkes 2011). The Yūgen aesthetic may also have infused embryologists' attempts to analyze, capture, and suture the time of embryonic development—in particular, their turn to film technologies.[22]

Wax models, serial sections, and mathematical equations were all part of a range of techniques used to grasp the course of embryonic development. "One could say that such analysis animates biological science: the merely observable regularities of life seem lifeless until the breath of formal, intelligible, analytic organization animates them as theory" (Kelty and Landecker 2004, 34). By the end of the nineteenth century, still another kind of animation was shaping the embryo, serving as a teaching aid in the early years of experimental embryology, only to be exiled to the role of whimsical decorations at the margins of scientific discussions in the years between the 1940s and the 1990s (Baedke 2013). I refer to embryo films and the still and animated images that sprang from them.

The turn to filmic animation was part of a broader attempt to explain organismic behavior by exploring its movement in three dimensions, whether on the level of the individual or of the aggregate. As I have described elsewhere, researchers at the Strangeways Research Laboratory in the 1920s drew on photography, cinemaphotomicrography (or microcinematography), and film to animate the embryos they observed (Squier 2004). Working with George Canti, who had been trained by Alexis Carrel in the new technique of tissue culture at Rockefeller University, Waddington and his fellow researchers learned to use stop-action time-lapse imaging and, later, motion picture film to depict embryonic development in process. Although he understood how shocking this new biological direction could be for scientists used to seeing their subjects fixed on the microscope plate and was concerned about the potential for distortion inherent in time-lapse animations, which "exaggerate the speed with which these movements are carried out," Waddington still acknowledged that this new technique could provide a useful counterweight to experimentalists' "tendency to envisage cells in terms of the static pictures presented by ordinary microscope preparations" (quoted in Kelty and Landecker 2004, 43).

Film technology made it possible to intervene in the time frame by which life unfolded (making it possible to speed it up or slow it down, or even reverse it); to look closely at the very moment of death; even to reconceptualize biological time. Death could even be reversed as the film was run backward, like T.S.P. Strangeways's revivification of a bit of sausage meat through tissue culture (Kelty and Landecker 2004, 43; Squier 2004). While film technology kindled an interest in the moment of cell death, it also redirected biological theory by introducing a new set of questions and a new set of experimental techniques: movement and multidimensionality now came to the fore, in contrast to the long tradition of fixing, stilling, de-animating, or freezing life to enable the scientific gaze (Kelty and Landecker 2004).

As embryology incorporated technologies ranging from hand lenses and microscopes to disciplinary frames drawn from anthropology, geology, and archaeology, the bulky materiality of the three-dimensional embryo model—so easy for a member of the public to understand, but to scientists less satisfyingly precise—gave way to two-dimensional visual representations. Researchers debated whether it was more appropriate to consider development in terms of "stages" or of "horizons." George L. Streeter, an embryologist at the Carnegie Institution of Washington and a student of Wilhelm His, argued that it was more accurate to use the concept of horizons, borrowed from geology and archaeology, "to emphasize the importance of thinking of the embryo as a living organism, which in its time takes on many guises, always progressing from the smaller and simpler to the larger and more complex" (Streeter 1942).[23] In turning to the discipline of geology to find a model for thinking about the embryo, Streeter anticipates Waddington's visual vernacular for the second version of the epigenetic landscape, whose hill on which the embryo or egg cell perches is etched by contour lines that recall nineteenth-century depictions of the geological strata in Scotland, familiar to Waddington from his training as a geologist. These contour ridges serve as another kind of animation, albeit one that takes place over geologic time. We might call it the animation of latency.

Lynn Morgan has written eloquently about the collaboration among pregnant (or once pregnant) women, researchers, and doctors that led to the creation of the Carnegie Embryo Collection, the compilation of "standardized" embryos held at the Carnegie Institution in Washington, DC, that made possible the process of developmental staging. As embryologists drew on multiple strategies to find embryos, soliciting them from physicians, hospitals, or the patients themselves, they were presented with multiple potential frames for the story their embryo sequence would tell. And while their research also focused on finding the causes of pathologies, as they compiled the so-called Carnegie Stages they engaged in the process of *unthinking* such alternatives to the normate origin story, among them the possibilities of miscarriage or teratoma (tumors that occur when pluripotent or embryonic cells develop atypically). Ultimately, Morgan argues, the Carnegie Embryo Collection reflected the "embryological view of life" founded on four basic beliefs: human life begins at conception and proceeds in an orderly fashion to birth; an embryo is the only possible outcome of a human pregnancy; embryos are simply biological entities and their significant aspects are strictly biological rather than social, historical, or philosophical; and embryological knowledge gives us access to the true "facts of life" (Morgan 2009, 11–12). As mediated over years by embryological knowledge, first descriptive and then experimental, this understanding of the embryo achieved its culmination in the standardized Carnegie embryos, so evidently expressing the very formal developmental regularity, discrete individuality, and implicit teleology that seem essential to what embryos are and can be. Whether serialized as models, drawings, plates, or in stop-motion photography or, later, film, the visual embryo was intended to communicate to viewers (scientific and lay) those foundational properties of embryonic life.

I have been describing the development from descriptive embryology to modern experimental embryology that provides the foundation for the central figure in the second version of the epigenetic landscape: an embryo rendered visible, subject to serial representation (in images that were first still and later moving), and accessible to intervallic visual analysis. Like Waddington's poem about "bubble blastula," however, my narrative up to this point has omitted the broader context in which this embryo is situated. To rectify that omission, I conclude the chapter by sketching two different perspectives on the embryo in the epigenetic landscape—one close to its creator, and the other farther afield—that can enable us to appreciate the full potential this image can have to illuminate biological development.

SCALED

As I remarked early in this chapter, Waddington's accomplishment in formulating the field of epigenetics, a synthesis of embryology and genetics, was to scale the embryo. He believed that time is an essential element to any biological understanding of development. Just as the changing lenses of photography present different spatial scales, from close-up microscopy to photograph to the panoramic lens of cinema, so he felt that biological life must be understood on multiple temporal scales. To truly grasp a living being, he argued, we must be able to think of it as "affected by at least three different types of temporal change, all going on simultaneously and continuously":

> These three time-elements in the biological picture differ in scale. On the largest scale is evolution; any living thing must be thought of as the product of a long line of ancestors and itself the potential ancestor of a line of descendants. On the medium scale an animal or plant must be thought of as something which has a life history. It is not enough to see that horse pulling a cart past the window as the good working horse it is today; the picture must also include the minute fertilized egg, the embryo in its mother's womb, and the broken-down nag it will eventually become. Finally, on the shortest time-scale, a living thing keeps itself going only by a rapid turnover of energy or chemical change; it takes in and digests food, it breathes, and so on. (Waddington [2014] 1957, loc. 200)

In making this distinction among the time scale of evolution, the life history of an animal or plant, and the minute-by-minute processes of breathing and digesting food with which a living thing keeps itself going, Waddington was differentiating not only between time scales but also, significantly, among the processes that characterize each scale: homeorhesis, the continuation of a stable developmental flow; homeostasis; and evolution. Moreover, scalar relations link the individual embryo to a population (of embryos) that precede it and follow it. Thus, as Waddington developed it, the epigenetic landscape depicted both the specificity of embryological development (at the scale of the fertilized embryo) and what he called "the full biological picture" made visible by the mediation of the microscope, as well as by statistics. The ongoing metabolic and physiological processes of development over the life span connect the embryo both to the "broken-down nag it will eventually become" and to the population from which they both have emerged and that they constitute in turn. From Waddington's perspective, the epigenetic landscape incorporated three time scales simultaneously: the time of embryonic development, the time of an organism's life span, and evolutionary time. Such an approach to the embryo is inherently process-oriented rather than teleological, because it attends not only to regularity but also to the departure from it and is concerned less with the individual than with the relationships and processes that continuously constitute being, in its multiple changing forms, in time.

RHYTHMIC

In conclusion, let us zoom out from that tight focus on its creator to the cultural and historical context within which Waddington's iconic image of the epigenetic landscape came into being. We have seen that different aspects of biological development were salient for Waddington depending on the context in which he was formulating them: with mathematicians and physicists at Serbelloni, he reluctantly acceded to their focus on precision and prediction, while in the artistic milieu in which he drafted *Behind Appearances*, he perceived connections and hazarded conclusions that transcended disciplinary categories. As this book was going to press, I came across a recently translated work by the historian of science Janina Wellmann that details, a century before Waddington and in another country, the broader context that, I have been arguing, is crucial to our understanding of biological development.

The Form of Becoming: Embryology and the Epistemology of Rhythm, 1760–1830, situates the scientific field of embryology within the emergence of a new "episteme of rhythm" across a range of artistic, philosophical, and scientific fields ranging from poetics, music theory, philosophy, and dance to fencing, military training, physiology, botany, and embryology. As Wellmann (2017, 18) describes it, this new way of thinking about organic life was "based on the core elements of rhythm—repetition, variation, regularity, period, modification, alternation, [and] relation." Integral to this new episteme was the use of "serial instructional graphics": visual images of bodily movement, in serial relation, rhythmically arrayed. Whether the images figured fencers moving through series of positions, soldiers engaged in a weapons drill, fishermen knotting a net, or a leaf unfurling, the pauses between images were as important as—indeed, more important than—the images themselves at conveying the emergent nature of life in time.[24]

Wellmann's argument that "rhythm mediates between the spheres of biology and culture" does not rely on "the trope of transfer."[25] Instead, she understands culture and science both as subject to the rules by which organic life is governed, whose core principle is rhythm. And it was this widespread, transcultural attention to rhythm that led to the emergence of experimental embryology, with its dedication to uncovering the epigenetic processes through the creation of the developmental series. This is exemplified, she argues, in the way the mode of working with series of embryo images changed, from the size-graded displays of Sömmerring to the plates of Pander and von Baer, whose series show not a gradual increase in size but instead show a change in form. The quality of this new kind of series that focused the investigation on how embryos changed, or "folded into life," was not the images themselves but what happened between them. "In the relationship of forms within the series," Wellmann points out, "the interval or gap between the pictures acquires constitutive force. It is the connection between the shown and the not shown, the bridge between fullness and emptiness, that generates change, and the spatial alternation of image and gap produces a new ordering of time. That order, the principle of the series, is its rhythm."

While Wellmann's dense, scholarly, and ambitious study is sure to provoke much discussion in the history of science community, its particular significance to me is methodological. Although it covers an era a century earlier than the one I address, and in another country, it shares my conviction that attention to the macro scale is essential to a full understanding of the development of embryology. Moreover, it offers a close-up look at the analytical affordances such a scaled-out view of development can provide. We will see in the next chapter what such a shift can bring to our understanding of the conceptual and methodological reach of epigenetic landscape.

CHAPTER 4

The Graphic Embryo

Chris Kelty and Hannah Landecker's valuable narrative of the role of film technology in the technical advances of embryology progresses from the fixed stained preparations under the microscopes, through micro-cinematography and time-lapse imaging, to the digital platform enabling the "mathematical formalization of biology" and, finally, its computer animation.[1] Yet it omits one crucial step of embryo animation: the representational strategy, discourse, and aesthetic that brought laboratory findings into the social realm. I am speaking here of cartoons and comics. As I discussed at length in *Liminal Lives* (Squier 2004), the researchers at Strangeways did not segregate their scientific from their artistic endeavors. They wrote poems, drew cartoons, and aspired to make films of the developing embryos they studied that would have the catchiness and crowd appeal of popular motion pictures such as *Steamboat Willie*. Although Kelty and Landecker (2004, 41) argue that time-lapse photography made it possible not merely to capture death but also to undo it, so that "the dead cell could be, like a cartoon character flattened by a truck, animated back to life, backward, to determine the exact cause of dying," they do not fully explore the meaning of the cartoon in that process. Perhaps they shift so quickly to the more serious, powerful, and authorized modes of film and digital media rather than exploring the meaning of that "static cartoon" and the flattened cartoon character because comics and cartoons have been often dismissed as sources of knowledge, assumed to be the simple medium of childhood and stigmatized by notions of developmental delay (Squier 2008).

In this chapter, I begin with the moment at which the film "cuts between a hand drawing a diagram of a cell on the board, a static cartoon with its organelles labeled, and time-lapse sequences of living cells, writing and crawling across the field of view." Linked by its rhythmic seriality to the embryo imaging before it, which exploited the pause or gap between images to inquire into the very processes of morphological change, the cartoon, too, has been instrumental in shaping our understanding of cells as capable of motion and change. In fact, it was not only photography and film that enabled embryologists to capture the developing embryo and show it to an interested public, garnering support for the embryological mission. Another medium by the twentieth century had also acquired a wide popular audience and was thus a powerful source of communication: cartoons and comics. Arcane as their own scientific research may have seemed to the public, the tissue culturists at Strangeways were nonetheless well aware of the publicity power of Mickey Mouse. The researcher Petar Martinovitch revealed this in the poem he wrote about their laboratory work; he recommends that the tissue culturists reach out to their potential public by imitating "Mickey Mouse and his gait" and telling their story in a cartoon motion picture (Squier 2004, 85).

The immortal mouse was introduced to the international public in *Steamboat Willie*. This animated cartoon by Walter Disney (as he was described in *Variety*) impressed the reviewer with its "high order of cartoon ingenuity combined with sound effects. . . . With most of the animated cartoons qualifying as a pain in the neck, it's a signal tribute to this particular one. If the same combination of talent should turn out a series as good as 'Steamboat Willie,' they should find a wide market."[2] This cartoon, with its combination of ingenuity and marketability, has its origins in the rise of the comic strip. Animated cartoons originated as drawn images. From the pioneering work of the Frenchman Émile Cohl and his matchstick drawings through the work of the American Winsor McCay (whose "Little Nemo in Slumberland" is a comics classic) to the Australian Pat Sullivan, animators brought the drawn line, time, and space together to create vivid life (Halas and Manvell 1968, 26–27). But comics had a longer history, beginning in Great Britain in the 1870s, with a pullout supplement to the *Weekly Budget*. When the *Budget*'s publisher, James Henderson, realized that there was an audience for this new medium, he created the first comic weekly, "Funny Folks." From the deposit of the first comic in 1874 and the color pages of the highly popular "Puck" to the "Mickey Mouse Weekly (1936–1955) . . . the first comic to be printed in full colour photogravure," scholars can trace the growing popularity of the medium, thanks to the excellent *British Comics Collection*.[3]

By the end of the nineteenth century, the comics publishing boom had begun: the long-running "Comic Cuts" (1890–1953) emerged, created by the publishing mogul Alfred Harmsworth (Lord Northcliffe). With the slogan "Amusing without Being Vulgar," this comic used the marketing strategy of competitive pricing to achieve success in reaching the working-class public; at one halfpenny, it cost half the price of previous comics. Northcliffe had many critics of his publishing strategy, not least because he held a position on education that differed from that of the reigning elite: "'Explain, Simplify, Clarify' was his motto. Yet his combination of segmentation, seriality, and populist humor was a winning one, and soon other publishers began to compete with Comic Cuts, producing similar comics with names like 'Funny Cuts' (1890–1920)."[4] From a simple addition to the daily newspaper aimed at attracting broader readership, the new medium of comics had developed into a highly popular standalone publication giving voice to the voiceless in both the United States and Great Britain.

This chapter returns to the scalar embryo of modern experimental embryology to consider how it was newly mediated and animated in comics and cartoons. Before I go on to explore this point at length, however, I want to consider whether and how this representational strategy for the embryo differs from the second version of the epigenetic landscape—that abstract image of the ball on a hill that was the subject of chapter 3. *What is animated in the embryo by the turn to comics?* According to Evelyn Fox Keller (2002b), three major technical advances in the twentieth century led to a renewed interest in embryology, the first one enabling new "modes of intervening"; the second providing new ways to look at the findings those interventions produced; and the third providing new ways to represent the information visually. Keller is talking here about the remarkable innovations into biological microscopy made possible by the wide varieties of methods for DNA sequencing and the new capacities that computers provide to analyze and exhibit findings (Keller 2003, loc. 2318).

I argue that comics offers another technical advance in embryo mediation. This medium enables us to intervene differently in the meaning of an embryo, to consider it in a wider context, and to represent its meanings more widely across a broad social field. The turn to comics to communicate about embryos thus illustrates what Keller has called "explanatory pluralism." This quality, Keller (2003, loc. 3159) suggests, "is now not simply a reflection of differences in epistemological cultures but a positive virtue in itself, representing our best chance of coming to terms with the world around us." The verbal/visual comic genre I call graphic embryos demonstrates a multidimensional engagement with development that expands both our sense of biological and social possibilities and valuable aspects of the epigenetic landscape by remediating the traditional images of embryos. Representing and refiguring the embryo, works of graphic medicine circulate in public and clinical spaces as part of a network of *remediation*; by moving the attention to biological development into the social realm, they make possible a reparative engagement with the practices and personnel of contemporary medicine.

I will define "graphic medicine" in a moment, but first let me explain how I am using "remediation." The term carries resonances not only of landscape architecture, urban planning, and pedagogical research (it figures in all of those fields) but also of media studies. In ecologically minded landscape architecture and regional planning, the term refers to measures to clean up a brownfield site or to return polluted water to clarity; in pedagogy, it refers to methods that enable students to catch up with their faster peers. Yet when they repurposed the term "remediation," Jay Bolter and Richard Grusin (1996) were interested in exploring the way that any new media has a tendency to take on, revamp, and reframe the media that come before it. I draw on all of these denotations and connotations in the discussion that follows. I mean "remediation" in Bolter and Grusin's sense, as a mode of refashioning earlier forms of media, as well as in the sense the verb holds in everyday use (to settle a dispute, bring about a compromise, intercede), and in the sense of its cousin, the noun *remedy* (from the Latin *remediare*): a treatment, antidote, medicine, corrective, repair, or renewal.[5] The specific genre I call the graphic embryo illustrates all of those meanings of remediation: not only does it reframe and counter the earlier media technologies that have been used to make embryos visible (and the pregnant women invisible) so that we can track biological development, but graphic embryos also expose the conceptual mediations and the institutional and theoretical frameworks that have shaped, and narrowed, the context within which we engage with the embryo. As I explore in this chapter, some very positive feminist possibilities are generated when we reimagine, and remediate, the embryo.[6]

But what do I mean by "graphic medicine"? The term was coined by the cartoonist and physician Ian Williams to connote "the role that comics can play in the study and delivery of health care."[7] He intended the term to describe comics that address issues of illness, medical treatment, caregiving, and disability from the perspectives of patients, family members, caregivers, health-care professionals, and friends. In this comic genre, which is increasingly being used to address problems in institutional medicine, "remediation" takes on its fundamental meaning: to remedy.

The problem these comics address is as old as the alienation from the whole organism that drove Wilhelm His to engage in freehand modeling of embryos and as new as evidence-based medicine. Not only has molecular medicine given us a new scale for engaging with patient care, but with the incorporation of medical genetics into the arsenal of clinical medicine, another scale shift has occurred and an important distinction is being obscured: that between the individual patient and the population in the aggregate (Keller 2010). This conflation of terms is evident in the move for evidence-based medicine, which began as a push instituted by physicians and inspired by the Rand Corporation to incorporate the epidemiological principles of population-based medicine into protocols for individual patients' treatment and medical education and has now assumed central importance in the field of institutional medicine.[8]

To this field of evidence-based medicine, with its emphasis on statistically based diagnostics, treatment, and prognosis, graphic medicine brings the force of individual experience in all of its ambiguity and diversity. To medical education, it brings the opportunity for reflexivity. Physicians, who now learn in a number of medical schools to create comics about their own experiences, and patients, who are now creating their own "graphic pathographies" about their experiences of illness and medical treatment, have the opportunity to access their subjective experiences of providing and receiving health care by reading and creating works of graphic medicine. To medical ethics, graphic medicine brings an alternative, embodied, visual mode of expression that can enhance, and at times even disrupt, the rationalist, internalist emphasis of conventional bioethics discourse. To the medical humanities and its allied field of literature and medicine, graphic medicine brings its unconventional communicative strategies, disrupting the conventions and value hierarchies of canonical literature.

Within the graphic medicine genre, the specific comics I call graphic embryos share the attention to communication between scientists and the lay public that originally led to the long tradition of embryo modeling and embryo imaging. We have seen that Samuel Thomas von Sömmerring employed the serial arrangement of embryos in rows, with both the size of the embryo and its time in utero increasing steadily from the top left of the image to the bottom right of the image.[9] Josef Benedikt Kuriger did the same with the wax reliefs he created based on them, while Karl Ernst von Baer used the visual conventions of arrows and dotted lines to indicate developmental progression and directionality.[10] The comics medium carries forward this embryological interest in serial and three-dimensional models in several different ways. First, the sequence of panel placements in comics requires that the reader's eyes move from top-left to bottom-right across the page, unless indicated by an arrow or dots that guide it in a different direction. Then, like von Baer, comics also incorporate arrows, zip ribbons, lines of various thicknesses, and stylized motion lines to convey movement (McCloud 1994, 110–11). And finally, like the graphic instructional diagrams of the eighteenth century, comics achieve their effects by rhythmically alternating images of human or animal bodies with the spaces between them, spaces that require of the viewer the work of analysis.

Graphic embryos address the same problem that motivated His: the estrangement from a holistic view of developmental processes that led His to turn to embryo modeling to remedy (Hopwood 1999). But they also remedy the problem His faced in his day: "One had become estranged from the form of the whole embryo by contemplating isolated sections" (Hopwood 1999, 485). A similar technologically based focus on specialties and organ systems has led to the isolation of the physician or other health-care worker from the embodied experience of the whole sick person in contemporary biomedicine. His called for a "doubly embodied knowledge"—the kind of knowledge produced when scientists "work not only with their brain and eyes but also with their hands" (482). "Working on the same object *from different aspects* over a period of weeks built up a three-dimensional mental image. There was no substitute for this experience. . . . the modeler

His agreed with the artist Ecker that 'the pictures in the memory that have once made their way through the hand stick much more firmly in the head'" (483). Yet even more than His's embryo models, works of graphic medicine facilitate an encounter with the whole patient. Comics retains this dynamic in the relation between the drawn line and the three-dimensional body with which the cartoonist necessarily works, a body whose gestural dimensions, we will learn, are both affective and multiple. This connection of hand, head, and heart explains the powerful effect on both cartoonist and reader that comics can provide through their engagement with embodied experience (Chute 2016; Squier 2015).

Just as His's embryo models were part of a range of techniques used to "grasp" the embryo (including "plastic reconstruction, simple experiments, and . . . equations"), graphic embryos enhance, rather than displace, other ways to engage with embryonic development, from the medical to the social, political, and even spiritual.[11] Embryologists responded to His's work, gradually acceding to his belief that "in the microtome age they must reconstruct complex forms in three dimensions" (Hopwood 1999, 486). Increasing numbers of contemporary physicians and nurses welcome the multidimensional perspective on illness and care exemplified by graphic embryos and the broader comics genre of graphic medicine. In fact, as once embryologists inspired by His turned to making their own models of embryos, now health-care workers have even begun to create their own comics. In this they join not only cartoonists but also patients, family members, and caregivers, with effects that I consider.

Yet as graphic embryos draw on the representational conventions by which the embryo is mediated in the field of embryology—seriality, three-dimensional modeling, and the deliberate decision to use representations of embryos to connect scientists to the lay public—they do so, frequently explicitly, to reshape them. Thus, they differ from the previous tradition of embryo images by incorporating the principles of perspectival multiplicity, representational dissonance, and embodied knowledge. They put the embryo in a broader context, portraying not merely its stages or developmentally arranged sections, but also its personal, social, embodied and institutional meanings, which may be in tension. And they do so in space and time, combining verbal with the visual, the perspective of writer with that of cartoonist, the aural with the tactile. The loop between the mind of the cartoonist, as he or she both

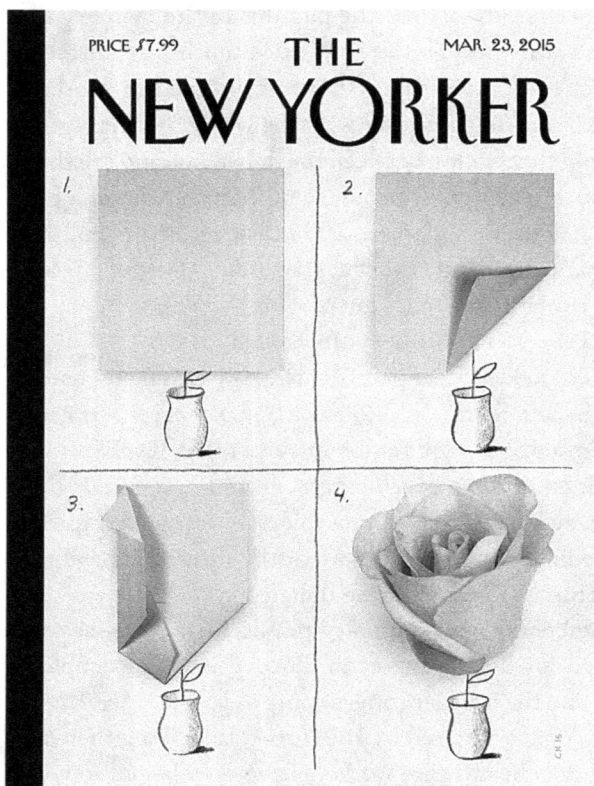

FIGURE 4.1 Christopher Niemann's "A Rose" (2015). This *New Yorker* magazine cover was a comic that, when viewed online, became an animation. The Post-It note folds and refolds to become, at last, a blooming rose.

draws and *draws in* the morphological changes from panel to panel, and the mind of the reader, as he or she both reads and consumes the comic, produces the multidimensionality of a third space.[12]

Christopher Niemann's cover for the *New Yorker* magazine (figure 4.1) captures the crossover between the embryological and the comics aesthetics, or between two-dimensional and three-dimensional, in its compact combination of seriality, rhythmic folding (or unfolding), and animation. Four numbered panels contain a pencil sketch of a vase with a stem, on top of which lies a pink Post-It in various stages of pleating. The panels are to be read from top left across to top right, and then from lower left across to lower right. In the online version of the *New Yorker*, this comic becomes an animation. Over the course of four panels, the pink Post-It that lies flat atop the sketched stem folds first a corner, then several paper petals, and finally becomes a living rose. As with all comics, the action in the *comic* is in the gutters—the space between those images of the gradually folded pink paper square—which are a site of agency that disappears with animation, as the rose comes into vivid life as by autopoeisis.

Niemann's animated comic illuminates the relation among the formal properties studied by embryologists, phenotypic expression and three-dimensional movement, and the representational conventions of comics. The embryologist studies the development from pluripotency to specificity, that embryonic path modeled by Waddington's epigenetic landscape: "the causal interactions between genes and their products, which bring the phenotype into being" (Waddington 1942). Catalyzed by the new attention to rhythmic seriality, embryologists use visual images to examine the pulsations and pauses of embryo folding. Moreover, just as epigenetics examined the specific morphological and behavioral traits of the embryo that give rise to the specific phenotype, the attention to traits—physiological as well as behavioral—is a central cartooning strategy.[13]

Comics as a medium builds its images through the winnowing out of visual possibilities, according to the veteran cartoonist Will Eisner. He explains the relationship among the drawn line, the physical manifestation of a character, and a character's behavioral repertoire in a taxonomy he presents in his study of the "Expressive Anatomy of Comics." Eisner's "micro-DICTIONARY of GESTURES" (figure 4.2), a one-page guide to the expression of the emotions through comic morphology, recalls the anatomical atlases and instructional graphics of the seventeenth to nineteenth centuries in the way it presents its subject visually in a sequence of different body shapes and gestures.

Perhaps because he is aware of the rhetorical difficulty of linking scientific taxonomies and the broader scientific project to artistic and even cartooning skills, Eisner (2004 [1985], 103) defends his attempt "(like da Vinci) to make a science of art":

> Formal or organized recorded communication began as visual communication. It is therefore not surprising that the artist can count on wide reader "reception" when a common gesture is limned so that it is

FIGURE 4.2 An excerpt from Will Eisner's "micro-DICTIONARY of GESTURES," a guide to the expression of emotions in comics (Eisner 2004 [1985], 102).

easily recognized. The skill (and science if you will) lies in the selection of the posture or gesture. In a print medium, unlike film or theater, the practitioner has to distill a hundred intermediate movements of which the gesture is comprised into one posture. This selected posture must convey nuances, support the dialogue, carry the thrust of the story, and deliver the message.

Eisner (2004 [1985], 102) argues that, in comics, "body posture and gesture occupy a position of primacy over text." Looping through the reader's own experience, the forms of posture and gesture not only internalize the emotions embodied in the comic frames, but also give an "auditory inflection" to the text—an internal experience of hearing the words spoken. Thus, body posture and gestures can intensify, affirm, contextualize, or even negate the meaning of the words within the speech bubbles, as does the fact that the speech is frequently hand-written.[14] This notion of a micro-DICTIONARY thus links one sort of expression to another, the affective unconscious to the complex process of gene expression.

Susan Oyama has described development as passing through "evolution's eye" to give rise to a unique individual. A similar hourglass distillation of expressive life figures in Eisner's theory of sequential art. In the development of a comic image, he explains, the "hundred intermediate movements" to which our experience gives rise are sifted, through artistic skill/science, until the cartoonist chooses a "common gesture" or a particular image from among the many thumbnailed. In the physical forms of Anger, Fear, Joy, Surprise, Deviousness, Threat, and Power, "these very simple abstractions of gestures and postures deal with external evidence of internal feelings," Eisner explains (Eisner 2004 [1985], 102). The plasticity of the expressive gestures in the micro-DICTIONARY suggests that the comic medium may be more flexible than print.

For example, the single word "Anger" in Eisner's "micro-DICTIONARY" offers seven different gestural expressions, or what we might think of as multiple comic alleles. Each internal feeling is expressed not just in one image but in a number of them, and even those are chosen from the "enormous bank of symbols we build up out of our experience" (Eisner 2004 [1985], 102). Any gesture, then, is one of many possible ways to express feelings, chosen from the wide sweep of expressive anatomy available to the cartoonist for her specific comic image. Eisner's phrasing recalls Waddington's insight that the specific genetic properties of any individual being are selected from the vast number of genes in the population reservoir and that, as epigenetics reveals, those genes are expressed differently based on the interaction of the environment with the developing individual.

Another crucial aspect of the representational conventions of comics is iconicity. Consider the iconic cartoon avatar, that quick sketch of a face, with dots for eyes, a vertical line for a nose, and a straight line for a mouth. If we return to Michel Serres for a moment, we can come to understand how iconicity embodies a crucial property of the embryo in the epigenetic landscape: its totipotency. Paying tribute to his friend the Belgian clear-line cartoonist Hergé, Serres explains that the appeal of Hergé's "Tintin" comics lies in the iconic face of their protagonist: "Each reader thrusts his own body into the straits left by this white absence and says to himself in evoking it: I am Tintin. The adventurer, in turn, whatever he is called, identifies for the same reason and participates in a thousand diverse individuals—from every class, ethnicity, culture, latitude—in the characters of this encyclopedia made up of ellipses and parabolas that make Hergé into the Jules Verne of the first human sciences" (Serres et al. 1997, 158). The properties Serres celebrates in Hergé's comics echo those of the epigenetic landscape: the transgenerational transmission of traits through embryonic development. Like the embryo at the peak of the epigenetic landscape, representative of "a thousand diverse individuals," Hergé's iconic cartoon character Tintin and the comic in which he figures also link pluripotency to transgenerational transmission. Hergé's comics were read "continuously for more than half a century by several generations, each rereading it as the following generation discovers it." As Serres recalls, "Trait by trait, [Hergé] thrilled our childhood" (155–56).

Serres's choice of the word "trait" has multiple valences, even beyond the epigenetic overtones we have already considered. In his native French, the term refers to the drawn line. In English, however, it conveys a particular temperament or mode of behavior, a character trait (MacDonald 1998, 2). Noting this conflation of the verbal and visual notion of trait, Amanda MacDonald (2) has argued that in Hergé's comics, "The entirely visible, graphic character, and the largely conceptual personage are . . . both made of traits." The drawn line and the verbally crafted character link the morphological aspect of Hergé's images with what we might call the behavioral aspects of his words. Indeed, the drawn line of comics links creator and reader/viewer both affectively and physiologically, with implications explored later. "Comics is a haptic form for both its creators and its readers," Hillary Chute has observed. In a study connecting the drawn nature of documentary war comics to the visceral impact of a centuries-long tradition of rendering artistic witness to trauma, Chute borrows from W. J. T. Mitchell's *What Do Pictures Want?* to argue that the process of "materializing" history onto the page combines the love and death drives as an enactment of desire (Chute 2016, 130). "Drawing itself," Mitchell writes, "the dragging or pulling of the drawing instrument, is the performance of a desire. Drawing draws us on. Drawing just is, quite literally, drawing or *a* drawing—a pulling or attracting force, and the trace of this force in the picture" (quoted in Chute 2016, 27). While I am persuaded by Chute that comics has certainly joined the tradition of documenting the death drive enacted in centuries of war making, I argue in this chapter that in its rhythmic instantiation of sequence, visual image, and pause or gap, comics even more thoroughly embodies the desire for life that triggers embryonic development. Comics thus renews and redirects toward greater flexibility, inclusiveness, and depth the long tradition of making embryos visible.

We can summarize the central qualities of comics as a medium, then: its attention to the expressive power of form and shape; its incorporation of a process of distillation from a vast population of potential images to a single individual image; and its use of the iconic image as having the greatest potential expressive reach and, through the act of drawing, the double-valenced attractive force of desire. But what is the methodological payoff for those of us interested in illuminating the epigenetic landscape, in making this heuristic use of the new medium of graphic narrative? If we consider graphic embryos part of the embryological history extending back to the serial images of His and Sömmerring, we discover earlier concepts about embryos reanimated in the formal and thematic properties of contemporary comics. For example, the iconic image of Tintin's blank face may recall the pluripotent egg cell at the top of the epigenetic landscape; comics' attention to form and shape may bring to mind the embryological study of morphology; and in the process of distillation by which cartoon gestures are selected, we might discover reverberations of the processes of inheritance, development, and differentiation that, to the first epigeneticists, explained the progression from genotype into phenotype (McCloud 1994, 45; Serres 1997). Yet there is more to it than that: in experimenting with the tropes and tools drawn from the medium of comics we can engage in the process of formulating a hypothesis and working it through visually, a conceptual and methodological act of reframing that illustrates how the epigenetic landscape can serve as a prompt to thinking even beyond the realm of biology.

Let us return to the specific case of the medical understanding of development, framed theoretically and practically by embryology, epigenetics, genetics, and, now, postgenomic medicine. The genre of graphic embryos offers multiple aspects, or perspectives, on embryonic development, and as it does so it enriches our embodied knowledge. In what follows, we will move through comics that reintegrate the perspectives of gender and disability into our understanding of embryonic development, beginning with "Bad Blastocyst"—a comic by Ruben Bolling that shares Waddington's concern with embryonic development, yet explores the wide variety of bioethical and social issues an embryo can generate—and concluding with a comic that recalls the epigenetic landscape at its most abstract and schematic: Richard McGuire's *Here* (2014).

"BAD BLASTOCYST"

The first embryo comic I ever encountered was "Bad Blastocyst" (figure 4.3).[15] I was writing about the controversy over stem cell research when I discovered this ten-panel, three-tier comic by the American political cartoonist Ruben Bolling and was impressed by its compact expression of the rhetorical pluripotentiality of stem cells and their mobilization to

FIGURE 4.3 "Bad Blastocyst," a comic that remediates medicine's instrumentalization of the embryo (Bolling [Ken Fisher] 2004).

support a wide range of incompatible arguments (Squier 2004). Later, I discussed it in my contribution to *Graphic Medicine Manifesto*, where I talked about how it could "fuel an ethics seminar in just ten panels" (Squier 2015). I can't seem to stay away from this touchstone comic. Now, reading it again, I think of it less in terms of ethics than as a response to the challenge posed by experimental medicine: balancing the well-being of an individual patient against the desire to generate new scientific knowledge and put it to use. Understanding it now as part of the genre of graphic embryos, I view "Bad Blastocyst" as a comic that *remediates* the way that medicine has created, understood, and instrumentalized the embryo.

Let's watch that remediation at work. The title panel of "Bad Blasto-cyst" offers us the familiar image of the blastocyst as viewed through the microscope. Embryologists, watching the embryo through the micro-scope, have taught us to see it as the detached object of scientific observa-tion. Yet in the same panel, the title text, in typical comic font, moves us from the world of science into the realm of popular culture, the home of another sort of media: pulp fiction and journalism. The question it poses positions the embryo as a particularly interesting form of popular dra-matic subject: "Was it bad? Or just misunderstood?" The next two panels introduce the basic plot: a cleaning woman at "Sal's Fertility Clinic N' Stuff" bumps a cabinet, dislodging a petri dish, which "falls on her head—killing her instantly!" Remediation continues through the first five pan-els of the comic, which offer two visual puns on "media": the petri dish in which the blastocyst grows and the montage of newspapers whose front pages record, in a range of journalistic modes, the act of homicide the "bad blastocyst" has committed. In those same panels, our perspective as readers changes: despite the scientific priming offered by the panel, with its microscopic image of the blastocyst, we quickly abandon that perspec-tive as we come to think of the blastocyst not as a scientific object but as a proto-human being. This occurs not only because of the blastocyst's sub-jectification in the title panel text, but also because the visual perspective in panels one and three aligns us with the blastocyst, as do the emanata in panel two. Central items in a cartoonist's toolkit, these droplet-shaped expressive marks reveal the blastocyst's response to the "Bonk!" collision in the same way that the cleaning woman's closed eyes reveal hers. The caption for panel four reinforces that shift in identification, from the "instantly dead" cleaning lady to the "tiny human blastocyst" that has survived "the incident."

Another mode of remediation appears in the next two tiers of panels, as various social institutions respond to "the incident" that has thrust the blastocyst into the media spotlight, asking new questions and giving new answers about the nature and meaning of human identity. When the lawyers become part of the story, the issue of capital punishment reconfigures the status of that microscopic life form; when research scientists enter the picture, questions of ethics come to the forefront. While the white-coated researchers argue that the blastocyst should be used in scientific research, the headline offers an opposing view: "Scientists call for Embryonic Execution." Finally, even the boundaries of human life are called into question, as the blastocyst is found to carry a genetic mutation marking it as "a retarded person," thus protecting it because, as the defense lawyer explains to the judge, "it would be *unethical* to execute it." In a final panel that references the philosopher Peter Singer's famous or infamous argument for a cognitively scaled assessment of the value of life, the scientists console themselves for losing the blastocyst as experimental subject, announcing instead, "Here! Experiment on this chimp! *That's ethical!*" But the ironic header has the last word, proclaiming (and undoing at the same time) "A Happy Ending!"

Recall, for a moment, the thick history of embryo mediation from the 1300s to the 1960s. "Bad Blastocyst" relies for its punch on an ironic remediation of many of those media spaces. The cartoon incorporates or refers to modes of visual mediation (microscope, newspaper, book, photocopy machine), as well as to institutional modes of mediation (law, science, ethics). Stories are essential to medicine, the psychiatrist Peter Kramer (2014) argues, because "stories are better at capturing a different kind of 'big picture.'" We particularly need the specific kind of stories that doctors call case histories, because they "convey what doctors see and hear." But we would be wrong if we merely saw "Bad Blastocyst" as another example of Kramer's argument that stories are an essential mode of medicine, viewing it as a narrative—a story—that enables the reader to navigate the social and cultural implications of the medical practice of embryo culture. Although I did put forth a version of this argument in *Liminal Lives*, I want to move beyond that position now.

Our unacknowledged hierarchy of text over image may explain why medicine, even narrative medicine, has been slow to acknowledge the contribution of comics. Consider that even as he argues that stories should inform clinical practice, Kramer builds his case in the language of painting: "Often the knowledge that informs clinical decisions emerges, like a pointillist image, from the coalescence of scattered information." Pointillism, we know, was Seurat's painterly innovation of applying small dots to his canvas so that they would blend into a recognizable image. What we may not know is that this technique anticipated the four-color printing process familiar from pulp comics themselves. Kramer's high-culture reference is undercut by its reference to the very printing technique linked to the quintessential comics format. As part of that pulp comic tradition, even if in newspaper greyscale, "Bad Blastocyst" enables in the reader the very same process of coalescing disparate information that Kramer attributes to case histories. Combining text and image, it can be read in both synchronic and diachronic fashion, as a linear narrative and as one nearly simultaneous compound image (Squier 2015, 46). Comic images arguably linger in the mind's eye, enabling—indeed, requiring—the reader to use multiple modes of mediation at the same time. So in "Bad Blastocyst" we see both scientifically and socially, with detachment and with affective charge, and we picture the blastocyst as both microscopic (the multi-celled product of in vitro culture) and macrosocial (subject of the international press).

Let us look at another example of how the graphic embryo remediates previous modes of mediating the embryo. *I Am Not These Feet*, the Finnish cartoonist Kaisa Leka's self-published comic, reveals both the cinematic aura of the embryo and the animating power of the cartoon medium for expressing congenital disabilities. Documenting the voluntary amputation of her congenitally arthritic feet, and her successful adaptation to a set of prostheses, Leka's comic makes explicit reference in its choice of avatars to the most famous animated cartoon characters, Mickey Mouse and Donald Duck, by representing herself as Kaisa the Mouse and her husband, Leka, as a duck.

Indeed, by imaging herself in utero as Mickey Mouse (figure 4.4), Leka pays tribute to the early encounter between time-lapse cinema and embryological research and then assertively reframes it. The comic recalls,

FIGURE 4.4 Mickey Mouse returns in Kaisa Leka's comic memoir about learning to live with a disability (Leka 2008).

though probably unintentionally, one of the earliest embryological references to the cartoon character, by Petar Martinovich, an embryologist in Honor Fell's laboratory at the Strangeways Research Hospital, in this poem I found in the Strangeways archive at the Wellcome Institute for the History of Medicine:

> The motto of the undertaking
> Of motion pictures making
> Is: you must tell a tale!
> So, of late, an attempt has been made
> By chromosome X and his mate,
> (the latter in its non-existant [*sic*] state),
> To imitate
> Mickey Mouse and his gait.
> If you like to see the show,—
> You need not spend your dough,—
> Come, and you will be told,
> Precisely, what you wish to know. (Squier 2004, 83)

Casting herself not as the suffering patient but as a Mickey Mouse–like "funny aminal" in the cartoon tradition, Leka at once references the embryological history that relied on such mediation to enforce the medicalization and normalization of the embryo and transcends it. Instead of linking her identity to her chronically arthritic feet she instead celebrates herself as a capable cartoonist, not only a hybrid of flesh and machine. As a final image reveals, by figuring her before a classroom of students, as a graphic designer and a comics artist, she is not a woman isolated and immobilized by disability but, rather a woman at work in the world, teaching others.

While graphic embryos such as "Bad Blastocyst" and *I Am Not These Feet* reference and refashion earlier ways the embryo has been mediated, technologically, conceptually, and institutionally, they do more than that. Just as the concept of remediation also contains notions of conflict resolution, environmental abatement, therapy, and cure, the embryo in comics extends beyond its iconic image in Bolling's comic or its cartoon avatar in Leka's. If we recall that the embryo is not just mediated and animated but also scaled, we will realize that the embryo invokes the human being at all stages, and scales, of life. Recall Waddington's assertion in *The Strategy of the Genes* that to truly grasp a living being, we must think of it as "affected by at least three different types of temporal change, all going on simultaneously and continuously." This means that we must think at the time scales of evolution, of an individual life, and of a cell engaging in respiration and metabolism. There is a spatial scale spread to this image, as well, as we have already learned from Waddington: It is not enough to see that horse pulling a cart past the window as the good working horse it is today; the picture must also include the minute fertilized egg, the embryo in its mother's womb, and the broken-down nag it will eventually become" (Waddington 1957, loc. 208).

Graphic embryos—comics that offer us visual and narrative mediations of the embryo—participate in all of those different spatial and temporal scales as they engage in expanded modes of remediation. When Hillary Chute suggested to Scott McCloud in an interview that comics can "rise above the landscape of time," his response had an uncannily Waddingtonian ring to it:

> That's the one thing that comics has that no other form has. In every other form of narrative that I know of, past, present, and future are not shown simultaneously—you're always in the now. And the future is something you can anticipate, and the past is something you can remember. And comics is the only form in which past, present, and

future are visible simultaneously. And in fact, if digital forms of comics were to allow us to put comics together in a fuller map, then that aspect and effect of comics would be amplified. (Chute 2014, 26)

As they widen the temporal and spatial scales of the graphic embryo, such comics expand the mode of remediation, exemplifying the extended use of the epigenetic landscape I explore in this volume.

David Small's stunning graphic memoir *Stitches* (2009) traces its protagonist's journey from a traumatic childhood of iatrogenic illness and familial neglect through psychoanalytic therapy to his artistic vocation. Central to this journey is a traumatic encounter with an embryo. Young David ventures from his grandfather's deathbed to a forbidden part of the hospital, where he encounters a display of *in vitro* embryos, lined up by age and size, according to Carnegie embryo exhibit standards. Floating at different levels in the glass jars and initially indistinct, the embryos gradually differentiate as David stares at them, until one embryo opens its eyes and glares at him malevolently. David runs away in fear, but in his imagination the embryo jumps from its jar and races after him down the long hospital corridor. Although David escapes to the safety of his parents, the embryo still haunts him, its malignant figure reappearing in his dreams. There, it swirls in a vortex along with the other objects David saw in "the X-rays of little kids' stomachs. The stuff they had swallowed like keys and pop beads . . . and cracker-jack prizes" (Small 2009, 28).

The back story here is significant: David has been injured by the X-ray treatments for sinus problems performed by his radiologist father as well as by his mother's wildly inadequate and distant mothering (a product of her mental illness). By picturing the embryo as something that could have been inside David himself, glimpsed through one of his father's X-rays rather than in a woman's womb, the panel sequence engages in multiple mediations and remediations. The size-graded embryo collection David sees in the sequence references the Carnegie embryo exhibits, a familiar feature of American life not only in hospitals but also in public exhibitions and at state fairs from the 1920s on, where its meanings were mediated through a range of practices from eugenics to sex education.[16] Following this indexical mediation, the embryo is transformed by David's own unconscious processes, and then finally, remediated in the psychoanalytic sessions through which David ultimately learns to release his anger, fear, and hurt. After these sessions, David returns to the hospital to revisit the embryo exhibit as an adult, and what had once seemed a malignant creature, face twisted in a baleful stare, now appears as merely an unborn fetus, its face unmarked because its life journey never began. Although historically these exhibits of embryos were intended to convey a population-scaled message, Small's mode of remediating the embryo in *Stitches* intervenes at the scale of an individual life, as his protagonist (and his readers) learn to see the embryo differently through the therapeutic mediation of psychoanalysis.

A further mode of embryo mediation is epistemological: we encounter it in the British cartoonist Paula Knight's graphic memoir about miscarriage, *X-Utero: A Cluster of Comics* (2013). This comic, part of a sequence created between 2011 and 2013 with the support of a grant from the Arts Council of London, reveals how different modes of attaining knowledge reveal different kinds of truth about the embryo. *X-Utero* was created as an online comic first and later self-published as a seventeen page minicomic. Yet this comic has had an impact on medical training that belies its noninstitutional origins; it has been adopted as part of the teaching material for genetic counseling students at Griffith University in Western Australia. Implicitly figuring itself as an embryonic form of intervention into how medicine addresses issues of miscarriage and childlessness, the mini-volume offers a "cluster of comics" with many potential developmental trajectories.

Moving from a comic about hospitalization for a suspected ectopic pregnancy (with the embryo shown as bright red in the fallopian tube), through a series of comics meditating on the inadequate ways to address miscarriage ("Lost vs. Loss," "It Wasn't Meant to Be," and "Failed") and failed pregnancy, to several powerful comics casting miscarriage and childlessness in a multigenerational frame, *X-Utero* testifies strongly to the painful ambiguities and uncertainties that go unexpressed in the "language concerning miscarriage and childlessness by medical professionals and others" (Knight 2013, n.p.).

The top two panels in "Lost vs. Loss" juxtapose an iconic image of a baby and a scientifically precise image of a sloughed-off embryo, revealing how graphic medicine can incorporate and communicate both scientific knowledge and personal uncertainty. The final four panels address the complex taxonomy of ignorance around women's health, revealing the

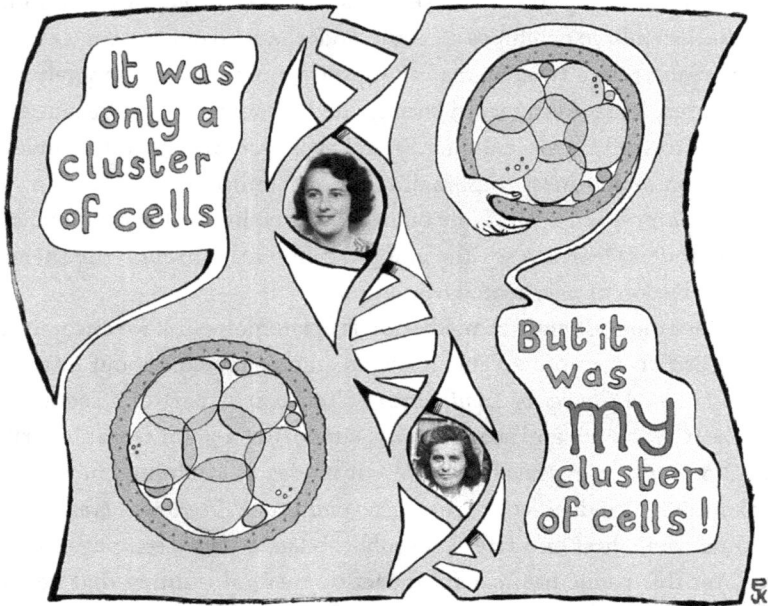

FIGURE 4.5 In "It Was Only a Cluster of Cells," Paula Knight (2013) uses juxtaposed images to explore the epigenetic origins of developmental loss and the complex transgenerational ties of biological relativity.

inherent ambivalence of the miscarriage experience, mingling personal certainty ("I did not lose this . . ."), with scientific knowledge ("My body rejected this . . . Carnegie Stage 10") and syntactic ambiguity (lost versus loss) that generates, even without meaning to, a sense of individual failure. By expressing the otherwise unexamined and unarticulated perspectives of women toward the embryo, alive or dead, Knight's work joins many other works of graphic medicine that reintegrate the gendered experiences exiled from formal medicine, not merely the feelings of patients and family members but their own forms of knowledge, expanding the scope of medicine in the process.[17]

Knight's comic "It Was Only a Cluster of Cells" returns us to an image of a blastocyst to explore the complex notion of biological relativity (figure 4.5). This concept, as formulated by the physiologist Denis Noble (2011), extended the relativity principle into biological development "by avoiding the assumption that there is a privileged scale at which biological functions are determined" (Noble 2011, 1). Noble asks, "Must higher level biological processes always be derivable from lower level data and mechanisms, as assumed by the idea that an organism is completely defined by its genome? Or are higher level properties necessarily also causes of lower level behaviour, involving actions and interactions both ways" (1)? Based on experimental and modeling work on heart cells, Noble proposes that both upward causation and downward causation shape biological life. This concept was later nuanced by the sociologist Sarah Franklin, who has expanded Noble's refusal of any privileged scale by moving the site of analysis from the biological into the social. Franklin argues that the new human relations created by in vitro fertilization also manifest a new form of biological relatedness (Franklin 2013).

Two kinds of truth about relatedness achieve simultaneous—indeed, co-constituted—articulation in "It Was Only a Cluster of Cells." The panel on the left pictures the embryo as detached (without relations): an object of scientific observation and manipulation. The title is framed in a speech bubble emerging from a classic microscopic image of the embryo. In the right panel, another mode of truth animates the embryo, as a speech bubble ending in a protective encircling hand affirms the embryo as *"my* cluster of cells." A middle panel links these two modes of truth, offering photographs of two women who—bearing in mind the different style, size, and coloring of the photographs—may be the author's grandmother and mother. While the double-helical structure of this panel clearly references emergent theories about the relationship between transgenerational genetic and epigenetic inheritance and miscarriage, it simultaneously affirms the other form of truth and the other mode of biological relativity that shapes our experience of a miscarriage: the understanding of developmental as something that entangles families across time and space, even through loss.

The cartoonist Diane Noomin's "Baby Talk: A Tale of 4 Miscarriages" (2012) expands on the feminist exploration of this theme of uncertain developmental outcomes, portraying the sequence from pregnancy to miscarriage as a series of embryo mediations ranging from scientific knowledge to folk fantasy, hallucinatory social commentary, and, ultimately, invisibility. Different modes of understanding the embryo coexist in the protagonist's mind. We begin with the pregnant Noomin following the growth of her pregnancy as she reads Lennart Nilsson's *A Child Is*

Born, a classic example of contemporary embryo visibility that relies on the performance of visual and cultural deception. (It is well known now that Nilsson's embryos were neither living nor representative of one embryo's progressive growth; indeed, even the framing and coloration of the images were the result of deliberate construction rather than the accurate portrayal of embryos "as they were in nature.") Lorraine Daston and Peter Galison's (2010) distinction among different epistemic virtues revealed through different modes of representation (truth-to-nature, subjectivity, and mechanical objectivity) illuminates Noomin's choices in this comic, although her vision succeeds theirs in acknowledging the possibility of a representation that ensures strong objectivity by presenting through voice and vision the woman's experience of her embryo.

Noomin imagines the growth of the embryo according to the taxonomy of normal embryo stages, but when pain intervenes and a deteriorating fibroid tumor threatens her pregnancy, her images of the embryo change. The medieval myth that a woman's dream can threaten the health of her fetus comes to the fore, as she dreams of her fetus as a lobster. Then, under morphine, she visualizes the embryo as the normate, gender-appropriate, ideal offspring, hallucinating embryonic versions of John F. Kennedy and Joyce Carol Oates. The juxtaposition of the Nilsson-mediated images of a staged and constructed embryo and the images appearing in her dreams and drug hallucinations relies rhetorically on the structural similarity between the image of a fetus in utero and a thought bubble. So Noomin's comic remediates the standard format of embryo imaging, interpreting it as the product not of mechanical scientific objectivity but, rather, of culturally and personally indexed subjectivity.

Rather than portraying the superstitious, scientifically constructed, or mass-mediated embryo, when the miscarriage finally occurs the focus of the comic shifts. What we *do not* see is crucial: we see neither the grisly photographs familiar from antiabortion billboards nor the serial displays of developing embryos like the ones studied by the historian Lynn Morgan in basement corridors at Mount Holyoke, where fetuses of various gestational ages exemplify normal development or testify to the ways development can go awry. In fact, we do not see the embryo at all. The focus has moved from the embryo to the direct perceptions of the woman who had been carrying it, and we see Noomin gazing into the toilet at an object outside the reader's view. "It looks like liver," she says. This example of the graphic embryo remediates the long tradition of scientific and culturally mediated visible embryos by removing it from public display entirely. Because it is not shown, the embryo is no longer subject to construction, interpretation, or manipulation by anyone other than the woman whose body once carried it.

Let us move back once again to consider the relationship between these two examples of the graphic embryo and the epigenetic context in which I place them. The works of Small, Noomin, and Knight can be read as responses to our postgenomic era, in which the bodies of pregnant women are understood as what Sarah Richardson (2015, 226–27) calls an epigenetic vector: a route by which the developing fetus encounters environmental influences that must be subjected to "processes of biomedicalization, optimization, and manipulation of life initiated by the twentieth-century molecular life sciences." Reductive, even determinist, these constructions of epigenetic medicine read the relations among grandmother, mother, child, and embryo as requiring risk management at the level of the individual. In contrast, these cartoonists are defiantly antireductionist and antideterministic. Small imagines a counternarrative of embryonic meaning fashioned through psychoanalysis; Knight reveals the multigenerational affective and social meanings of the embryo; and Noomin wrestles away the focus in a case of miscarriage from the assessed and imaged embryo to the frequently ignored and unvoiced subjectivity of the woman who carries it. These expanded representations of the embryo, from the narrowly individual to the more broadly cultural and relational, address and repair the narrowness of epigenetics as a mode of postgenomic research and medical practice.

One final way in which graphic embryos remediate is by shifting how the embryo is scaled and, in so doing, redrawing the boundaries of medicine itself. "Me and the Universe," a comic by the contemporary cartoonist Anders Nilsen, reveals the effects of thinking, and imaging, the graphic embryo at the scale of the landscape, the terrestrial environment, and even the planet and the universe. Nilsen's scroll comic offers an extension of the epigenetic landscape from the microscopic to the social and cultural.[18] First appearing in the *New York Times* on September 24, 2014, and published both in the print edition and its digital form, "Me and the Universe" explores the development of one human being in relation to the development of all life forms in the universe, throughout time and space.[19] Relying on the strategies of visual art and a range of narrative genres, from bildungsroman to science fiction and myth, it incorporates the productive ambiguities, complexities, and even contradictions that are frequently selected out of ecological and operations research approaches to development. Because it explores the factors that connect the embryo to both the individual and the population; relies on graphic gestures, as well as text, to convey the embodied and affective impact of a change of events; and strategically incorporates simultaneity and spatialization, as well as linear narrative, this comic actualizes the multiple time scales of the biological picture as Waddington understood them.

"Me and the Universe" combines comics conventions with biological ones. For example, it incorporates iconic images. Some of them come from evolutionary biology and embryology: the arrows that link fish to mouse to ape ("early prototype for the author"); the oak leaf and the acorn resonant of Darwin's arboreal images of inheritance; and the curled embryo, fetus, or baby. Other images mark the intersection of biology, astronomy, and anthropology, such as the evocative image of the corpse in repose, its internal space containing a starry galaxy of microbial life. The comic pictures the multiplicity of influences (or attractor basins) that shape embryonic development on the largest possible scale, thus reflecting Noble's expanded notion of biological relativity with its assertion that there is no privileged level of causation, as well as Franklin's extension of that notion. Representing the passage of human (life) time indexically, the comic connects the time scale of population to that of an individual life, as lines link specific points in the tree rings to specific developmental milestones: birth, sex, marriage, death, divorce. Nilsen's comic even incorporates the multiply scaled vision of the biological picture—at the scales of the microscopic, the individual, and the population—by moving from his birth to images of the earth as a globe ravaged by a "1,500 year cycle of disease, famine, and ecological upheaval" (Nilsen 2014).

Although Nilsen's narrative seems linear, it is actually recursive. Things repeat themselves, but with a difference, so that we experience not just evolution but its context: the nöosphere, in which what exists biologically and materially is intimately and reciprocally related to consciousness. Nilsen describes the "clouds of particles" appearing just after the Big Bang as "too agitated to connect in any meaningful way. (Similar to a 3-year period in my late adolescence.)" Later, a drive through the desert under the starry sky prompts him to realize, "This is all there is, I should try to appreciate it." This insight returns with a reflexive difference in the conclusion of the comic. When "the microscopic fragment that used to be you [the reader]," which is now lodged on a comet, crashes into "a neighboring star," there is a final "flicker of impact noted by sentient organism in solar system, who thinks: This is all there is. I should try to appreciate it."

Let me explain how the recursive form of Nilsen's narrative reflects the most recent understanding of how development proceeds. Since the turn of the twenty-first century, biology has returned to the model of the epigenetic landscape, finding it particularly useful because it can accommodate the big data produced by "high throughput gene sequencing" (Baedke 2013).[20] But in returning to that productive landscape, it does so with a difference. While Waddington's model (the epigenetic landscape) represented the determinative nature of development, demonstrating how canalization leads an individual to return to the normal (i.e., most evolutionarily beneficial) development course even when disrupted, recently scientists are discovering that the developmental process is neither so linear nor so determined. Cells can be reprogrammed—or, to put it in the logic of Waddington's model, the embryo in the landscape can be made to jump between channels or even to roll back up to the top of the hill.

This is the remarkable accomplishment of Nilsen's comic. Working between a timeline and a set of visual images, it represents simultaneously the developmental trajectories of one individual and those of the cosmos. It incorporates recursive developmental pathways by presenting the mermaid image of "one band [of humans] that begins its slow evolution back into water-dwelling creatures." And, in a particularly Waddingtonian accomplishment, it demonstrates the evolving course of consciousness in relation to the landscape of the universe, tracing the subtle change in a thought as it travels from the human being who is Anders Nilsen to the human being that is "you," the reader, and finally to an organism that is not necessarily human. Simply by replacing a comma with a period, the thought that Nilsen first ventured hesitantly is repeated decisively in the mind of that final sentient being: "This is all there is. I should try to appreciate it."

I have given this extensive reading of Nilsen's scroll comic because it speaks to trends in contemporary biology that have incorporated, extended, and reframed the epigenetic landscape.[21] Developmental biology in the era of evo-devo-eco is being framed within a far broader scale, as the development of life forms (including the human) is now being understood in the context of processes and relationships that cross the boundary between the organic and the inorganic. Surrounding the two tree slices, with their indexical points of importance to human life, Nilsen represents the products of human industrial production: the plastic retainer, the Lakota arrowhead, and the shell casing of a U.S. government-issue musket. Indeed, the very time scale of a human life is recalibrated so that it stretches from the origin of the universe ("before which I did not exist") through the present (more or less) to the future. Consciousness, too, is redefined as trans-individual, as the author's mind registers and then communicates a message not only to the reader but also to a conscious, but perhaps nonhuman, being in another solar system.

Yet in interweaving stardust, microbial life, trees, fish, and animals; representing life at the scale of the human, the planetary, and the cosmic; and incorporating objects of organic, inorganic, and indeterminate materiality, Nilsen's comic moves us beyond the Anthropocene, with its focus on the human—and, arguably, the Western capitalist human, at that.[22] Instead, Nilsen challenges us to think in terms of earth, life, and systems. While the title seems to center the "Me" that is Anders Nilsen, the comic actually works to destabilize that anthropocentrism and the biomedical model to which it is connected. Of course, when we resituate the individual human life in relation to the fate of the universe, we are stretching the category of medicine, too, drawing it beyond the individual, and even beyond the population and the human species, to all species.[23]

We have explored the metaphysical concepts that inspired Waddington's work. We turn now to a comic that enacts them, created by the cartoonist who coined the term graphic medicine: *The Bad Doctor*, by Ian Williams (figure 4.6). Just as the chick newly hatched from the world egg looks back at the embryologist who studies it, Williams's graphic novel turns its interrogating eye back to the doctor. The comic takes as its central theme the reflexive interrelationship between self and environment that Waddington expressed in his obsession with the ouroboros and yet evaded in his talk at Serbelloni. We can see this expressed by the image on its title page: the ouroboros, the snake swallowing its tail. This allusion

The BAD DOCTOR

IAN WILLIAMS

FIGURE 4.6 The ouroboros reappears in *Bad Doctor* (2014), Ian Williams's memoir of medicine and OCD, signaling the reflexive relationship among doctor, patient, and the world in which they both live.

to cybernetic reflexivity is not merely metaphysical or theoretical for Williams. He is a physician himself and has taken the project of remediating medicine into the public sphere, contributing a regular comic-strip series, "Sick Notes," to the British newspaper *The Guardian*.[24]

The Bad Doctor concerns the day-to-day experiences of the Welsh doctor Iwan Jones as he deals with patients whose medical concerns run the gamut from psychiatry to geriatrics. A second, deeper plot line concerns the doctor's own struggle with obsessive-compulsive disorder (OCD). In his childhood, the OCD took the form of an elaborate set of encircling spells to keep his family safe (figure 4.7); in adolescence, as a preoccupation with Satanism and heavy metal music; and in adulthood, in recurrent, self-destructive fantasies and fears that those closest to him will be harmed.

Rather than his invisible disability, however, the main problem troubling Dr. Jones is the bad medical care he observes daily. Iatrogenic harm—harm caused by the activity of a physician, other health-care worker, or treatment—is a central theme of *The Bad Doctor*. Dr. Jones edits a patient's chart to remove another physician's inaccurate diagnosis of borderline personality disorder, caused when an insensitive comment "somewhere along the line . . . got turned into a diagnosis" (Williams 2014, 44). With his fellow doctor, he laments the harmful impact of the hospice nurse, "simpering Debbie," whose awful manner has incensed and driven away most of the terminally ill patients (114). Finally, Dr. Jones finds himself treating a patient haunted by the same spiraling obsessions and compulsions of OCD that so troubled him, and the issues come to a head. Obsessed by the fear that he will harm his nephews, the patient explains to Dr. Jones, "The more I hated myself, the more complex the obsessions got, and I became convinced that I was responsible for all kinds of calamities." Iatrogenic harm then follows: as he explains to Dr. Jones, when he told his mental-health worker about his obsessive anxieties, the mental-health worker "said that even though he 'knew I wasn't a risk to children,' as I'd 'voiced my worries' he felt 'duty bound' to report me to Social Services just to 'cover himself.'" Dr. Jones replies, "God help us," overwhelmed by the harm that doctors and other health-care workers inflict on patients. Dr. Jones makes the connection between his own illness and his duty to his patient. Spurred by this realization, he emerges from the veil of clinical distance to confide in his patient that he has had OCD from his adolescence through his medical school years. His act reframes doctoring not as harm, but as empathetic concern.

The tension among the different modes of knowledge shaping medicine—whether we call them subjective and objective, analogue and digital, or magical and the rational—persists even in the last chapter of the novel, signaled by the Tarot card image of the Wheel of Fortune with which it begins.[25] Dr. Jones takes his friend Arthur on a long bike ride to "a 'Rag Well' or 'Clootie Well,'" a clearing in the woods where bandages dipped in holy water are hung on trees in a mystical attempt at

FIGURE 4.7 Occult imagery preoccupies the young Iwan Jones in *Bad Doctor* (Williams 2014).

self-healing. Dr. Jones explains, "If you have an ailment you take a strip of cloth and dip it in the well. . . . Then you swab the affected body part and hang the rag on a tree. As the rag disintegrates . . . so the ailment will resolve" (Williams 2014, 219). "It looks like it's still regularly used," Arthur jokes. "People are obviously turning to folk magic to complement their NHS [National Health Service] care!" Dr. Jones's reply conveys the cost of such superstition as "our desperation for some sort of control over an indifferent universe. And the basis of many of our neuroses" (219). Then, as the two men stand there staring at the rag-draped trees, Dr. Jones announces, "I'm seeing a clinical psychologist. Should've done it years ago" (220).

I have mostly left the discussion of epigenetics to the side in my discussion of *The Bad Doctor*, preferring to highlight how the novel integrates the graphic embryo into its exploration of a physician's critique of the institution of medicine and his struggles with his own mental illness. Yet just as the novel reminds us of the metaphysical themes that led Waddington's own development from experimental embryology to epigenetics, in its focus on OCD *The Bad Doctor* addresses one of the areas currently spotlighted by epigenetic medicine. In the flashback episode triggered in Dr. Jones by his work with his obsessive patient, we learn just how smoothly contemporary epigenetic medicine's framing of the pregnant woman as vector for risk and environmental damage combines with the folk notion of malign maternal influence on fetal development (Richardson 2015).

In a chapter introduced by another full-page image of the ouroboros, we learn that Dr. Jones had been particularly tortured by his OCD during his wife's pregnancy. Occult images flood his mind when an ultrasound test reveals that she is pregnant with twins.[26] As the embryos continue to develop, Dr. Jones attempts to ward off evil by screening the music played during labor. The comic portrays his endless fears that Satanic music can damage the fetuses, in an embryo image referencing medieval representations of the pregnant Virgin (figure 4.8). Jones imagines that the backward messages hidden in Satanic music played over the radio could possibly injure the fetuses: "Does the music have to be actually heard or could the evil be transferred in radio waves as the songs are transmitted?" he wonders (Williams 2014, 169). The theme of epigenetic medicine also shapes the story of Dr. Jones's patient Mr. Crowley. His suicide reveals the bilateral defects of his irises that led to his stigmatization as "goat

FIGURE 4.8 Epigenetic medicine's view of the pregnant woman as a vector of illness intersects with the obsessive anxieties of Dr. Iwan Jones about the health of his wife's fetus (Williams 2014).

eyes," leading Dr. Jones to wonder how the hostile environment in which Mr. Crowley had lived throughout his life had reshaped his psychological development: "He'd been stigmatized since childhood. I wonder how much of his behavior was conditioned. . . . Maybe I could have done more. I might have saved him" (199).

Researchers now argue that racist health disparities and wartime famine can have inherited effects and that gender variants can result from the dynamic interaction of genetic and environmental influences (Sullivan 2013; Fausto-Sterling 2012). Behavioral geneticists even argue that psychiatric disorders such as OCD have an epigenetic aspect. Both psychological behaviorists and molecular behavioral geneticists have turned to twin studies, finding them important in determining the origin of the condition, as the authors of one recent review of the literature on obsessive-compulsive disorder explain. "A key result from these studies is that genetic factors are important for the manifestation of obsessive-compulsive behaviors and that non-genetic, non-shared environmental factors also have a considerable influence on the manifestations of OCD, presumably through epigenetic mechanisms" (Paul et al. 2014, 413). Moreover, they argue, "[The] findings from the twin studies imply that non-shared environmental factors may trigger the disorder" (413).

These behavioral geneticists seem to have moved beyond the gene-centered perspective, emphasizing instead multi-causality and complexity. Although a mode of lip service adherence to these qualities seems to coexist with the increasing interest in epigenetics (and even a reappearance of the epigenetic landscape), postgenomics may be "among other things, a mode of extending the project of genomics in the face of a restricted resource environment and increasing skepticism about genomics' prospects to provide concrete benefits for society and for human health" (Stevens and Richardson 2015, 5–6). Williams gives a wry rebuttal to these hyped epigenetic causes of behavioral genetics in the last two panels of *The Bad Doctor*, which reframe Dr. Jones's struggle with OCD in a broader sociohistorical context. As the two friends ride their bicycles home from the Clootie Well, they see a magpie on the road. "Shit! A lone magpie!" Dr. Jones says, "Quick! Find another!" This allusion to the old nursery rhyme reveals the folk understanding that sorrow and joy are inextricably, arithmetically, interwoven with the developmental process. Even as Dr. Jones affirms his intention to seek treatment for his OCD, the novel prompts us to understand that the medical model is not the only explanation for such behavior. Old thought styles persist, as Fleck tells us, even as new thought styles predominate:

One for sorrow,
Two for joy,
Three for a girl,
Four for a boy,
Five for silver,
Six for gold,
Seven for a secret,
Never to be told.

The point here is not that Williams chose to focus on OCD in *The Bad Doctor* because it had epigenetic connections, any more than that he chose to portray a twin pregnancy as the moment in which his character's OCD flared. Rather, as *The Bad Doctor* shows, the point is that those narrowly deterministic perspectives shape our sense of options and constrain the range of our possible responses. In contrast to epigenetic medicine, graphic medicine remediates. *The Bad Doctor* incorporates earlier representations of the embryo, epistemological as well as biomedical, precisely to reframe medical knowledge in a broader context and thus to imagine medicine done differently.

HERE

In his analysis of the iconography of illness, Williams considers how comics art portrays movement through iconography—that is, through a combination of "simplification and generalization" (Czerwiec et al. 2015, 118). We have been talking throughout this section about embryos that are depicted visually, but one recent comic conveys the process of development on a scale so abstract that the embryo itself disappears, leaving only the ongoing process of development. I am talking about Richard McGuire's tour de force of multidimensional cartooning, *Here* (2014).

Waddington had played with multidimensionality himself, entertaining the phase-space version of the epigenetic landscape before discarding it as too complex. I argue that comics as a genre recuperates that phase-space model, finally enabling the very complexity that Waddington both wished for and feared. The disciplinary hybridity is not merely an added bonus but precisely the strategy that has made it possible. "'Here' is the comic-book equivalent of a scientific breakthrough," wrote Luc Sante (2015) enthusiastically in his review for the *New York Times*. "In 'Here,' McGuire has introduced a third dimension to the flat page. He can poke holes in the space-time continuum simply by imposing frames that act as transtemporal windows into the larger frame that stands for the provisional now." *Here* shows us one space—the iconic and stripped-down corner of a house—through eons of time. Not much seems to happen—or, rather, many things happen, but in the most abstract and general of ways. In this simple corner of a rather generic dwelling space (featuring a hearth, a window, a lamp, a couch), children are born, old people die, men and women make love and fight, and walls are painted, stripped, and re-painted, built, rebuilt, demolished, and even seen through. The "here" of this comic is time seen as if from the fourth dimension—Waddingtonian time, if you will—which flows both forward and backward, happens at multiple scales, and is capable of multiple enfoldings, eruptions, and juxtapositions.

Scholars have already praised this complex and innovative comic, addressing its use of the medium's capacity to represent events both sequentially and simultaneously, as well as its incorporation of gutters and gutterless insets to enable a layered presentation of temporally disjunctive but spatially continuous moments (Klopmeier 2015). The digital e-book version of the comic even incorporates what McGuire calls "micro-animations," events he designed to be "a surprise . . . they don't happen every time. They're really tiny events" (Lohier 2014). (For me, this recalls Ronald George Canti's embryo films and the intense surprise invoked in early viewers by glimpsing movement at the microscopic level.) McGuire's online comic harnesses the power of random variation and multiplicity. "Everybody who sees the ebook," McGuire explains, "thinks . . . it resembles the book version. But if you play with it long enough, you start to click on the dates and the panels and that's enough of a clue that once you click on that one date in the upper left hand corner, it starts to open up the shuffling mechanism" (Lohier 2014). From there, the book unfolds palimpsestically, continually producing new patterns based on what McGuire marvels must be "an immense number of combinations that can happen" (Lohier 2014).

While the random and generative aspects of the e-book provide an interesting counterpoint to some of the ways that development is currently explored in biology using big data, I want to take a more retrospective turn in my approach to this final work of graphic fiction. I am interested in exploring how *Here* remediates an earlier experimental work of live-action animation and in so doing makes revolutionary use of the connection among embryology, animation, and comics.[27] The film I am concerned with is *Tango*, Zbig Rybczynski's experimental film from 1980 in which "thirty-six characters from different stages of life—representations of different times—interact in one room, moving in loops, observed by a static camera."[28] The plot premise of *Tango* recalls one of the earliest tropes in animated film: the bouncing ball. I remember singing along with television in my childhood—was it to Mighty Mouse cartoons?—by following the bouncing ball as it sprang from word to word of the lyrics. In this case, too, the animation was initially reliant on the mechanical process of crafting a material object. Wikipedia informs me that the device was invented by the Fleischer studio in Los Angeles in 1925. In its first iteration, it was no more animated or digital than the wax embryo models of Wilhelm His: "The effect was created by filming a long stick with a luminescent ball on the tip, which was physically 'bounced' across a screen of printed words by a studio employee. The movement was captured on high-contrast film that rendered the stick invisible. The ball would usually appear as white-on-black, though sometimes the ball and lyrics would be superimposed over (darkened) still drawings or photographs or even live-action footage. Later versions of the bouncing ball have used cel animation or digital effects."[29] A bouncing ball animates the action in *Tango*. It arcs in through the window of a small room, prompting a little boy to climb in after it and triggering the arrival of an increasing number of characters who enter and leave, cry and laugh, eat, have sex, and die, in no particular temporal order.[30]

Tango materializes Waddington's insight that the "biological picture" contains "at least three different types of temporal change, all going on simultaneously and continuously" (Waddington 1957, loc. 202). Like His's three-dimensional wax embryo models, which require artisanal work as well as embryological knowledge, the sheer manual and technical labor required to create this 8:10 minute film are central to its impact. Rybczynski has explained that error, friction, and time-consuming human and material wear-and-tear were involved in drawing and painting "about 16,000 cell-mattes" and making "several hundred thousand exposures on an optical printer" to create *Tango*.[31] *Tango* gives us an animated version of the epigenetic landscape, as the film explores the errors and misdirections to which development is subject. The child's ball is the totipotential egg, which triggers a torrent of different characters as it passes through the window and into the room (into life—or "here," if you will). In Rybczynski's telling, the material work to create the film echoed the contingency of the developmental process itself. "The miracle is that the negative got through the process with only minor damage, and I made less than one hundred mathematical mistakes out of several hundred thousand possibilities. In the final result, there are plenty of flaws: black lines are visible around humans, jitters caused by the instability of film material resulting from film perforation and elasticity of celluloid, changes of colour caused by the fluctuation in colour temperature of the projector bulb and, inevitably, dirt, grain and scratches" (Rybczynski 1997).

Like *Tango*, *Here* explores the remarkable multiplicity and error-filled generosity of biological life, on a scale extending from the microscopic through the human to the cosmic, and from the beginning of time to the end (or, as it seems, back to the beginning). Yet it does so more flexibly because of the very grammar of comics. As Sante (2015) observes, "Novels and movies are handicapped in their presentation of simultaneity by the fact that they are shackled to time themselves, in the physical unfolding of their narratives. The comic strip or graphic novel, however, is allowed to run free across space as well."

The temporal/spatial complexities are apparent in the comic's reception. To some critics *Here* seems to issue an environmental warning, while to others it seems to be a multigenerational family narrative. In either case, the epigenetic implications are invisible, implicit, but inescapable.

In *Here*, unlike in *Tango*, there is no boy to follow the embryonic bouncing ball through the window. Nor is there a visible embryo, although we do encounter multiple babies and children. Yet in its folded, reiterated, and rhythmic portrayal of human life on earth in that one corner, in the context of the numberless span of years, *Here* joins Nilsen's scroll comic in demonstrating how the course of an individual life can be framed simultaneously by the medical perspective on the human life course and the ecological perspective on the course of all life. Or, to return to concerns I raised earlier, it shows how the embryo and the landscape are both scaled and inextricably intermingled if you adopt the same strategy McGuire used in creating his comic: "make everything big, small, and make everything small, big" (Lohier 2014).

Let me conclude this chapter with two observations: one about comics, and one about the process of animation that unites comics and film as modes of embryo mediation. I have been arguing that works of graphic medicine are interventions into medical and health-care fields as much as they are into literature, the arts, and social practices. We do not know what the effect will really be on medical practice as more and more physicians create their own comics and as they incorporate the reading of graphic medicine into their training and treatment, although studies are already suggesting that medical students taught to make their own comics become more empathetic caregivers as doctors (Green 2015). We also do not know what will happen as patients, family members, and caregivers begin drawing on and incorporating the expressive anatomy of comics to explore the experiences of illness and medical treatment, but the explosion of works in graphic medicine does attest to the interest in the new platform for expression this medium offers. We do not know these things because the development that is graphic medicine incorporates the very characteristics for which the epigenetic landscape was abandoned as a biological model and that are at risk of being forgotten as it is once more being deployed in the life sciences: its analogical imprecision; its attention to the social, as well as the biological; its nonlinear, multiscale, multidirectional understanding of causation; and its resistance to disciplinary orthodoxies.[32]

Recall that in *The Strategy of the Genes*, Waddington (1957, loc. 557) observed, "[In] the study of development we are interested not only in the final state to which the system arrives, but also in the course by which it gets there." The best image of the epigenetic landscape, he acknowledged, would have been in a "phase space," or "a system containing many components [that] can be represented by a point in multidimensional space." While James Gleick does not take seriously the connection he notes between phase-space diagrams and so-called chaos comics, mathematicians' diagrams of phase-space motion, a connection worth exploring does exist between the multiple capacities of comics to embody potential experiences and spaces and the multidimensionality of phase space (Gleick 2011, 149). Indeed, in its capacity to "unflatten" experience, increasing rather than reducing the strings that tie us to our environment, as Nick Sousanis (2015, 135) has argued, the comic form makes us "better able to see these attachments not as constraints but as forces to harness."

The process of selection in which His and Corner participated as they came up with the developmental scale of the Carnegie Embryos was but one variant of that broader move to produce and stabilize embryo images to choose the "best" image of an embryo at a particular stage of growth. Cartoonists, too, select their images, just as they choose the words in their speech bubbles. Yet the forces flow both ways in this interaction. As they incorporate "unorthodox" images of embryos, cartoonists—because their mode of mediation extends beyond the scientific—counteract the standardization of experience that has threatened to reduce medicine to the application of statistically derived and population-based formulas for patient care. Thus, graphic embryos—comics that depict embryos—offer opportunities for readers of all sorts, inside and beyond the academy and the clinic, to reexamine the assumptions we make about what embryos are, how embryonic development occurs, what embryos can mean, and what health care should be. In their multidimensionality, graphic embryos may offer a more flexible and powerful model for engaging with the complex relations among embryology, genetics, and epigenetics. Not only do they enact the rhythmic, gestural, and artistically rich vision of life that motivated Waddington to create the epigenetic landscape in the first place, but in their multidimensional and temporally multidirectional aspect, they may requite his longing for a phase space representation of the epigenetic landscape.

We have been talking about a tango between graphic medicine and a medical understanding of the embryo. Recently, the cell biologist and animation artist Mhairi Towler, director and founder of Vivomotion, a company that has won awards for its science animations, has joined this effort, creating an animated version of the epigenetic landscape (figure 4.9). In a tradition dating back to the researchers at Strangeways Research Laboratory in the early twentieth century, EpiGeneSys, which describes itself as "an ambitious E[uropean] C[ommunity]–funded research initiative on epigenetics advancing towards systems biology," called on animators

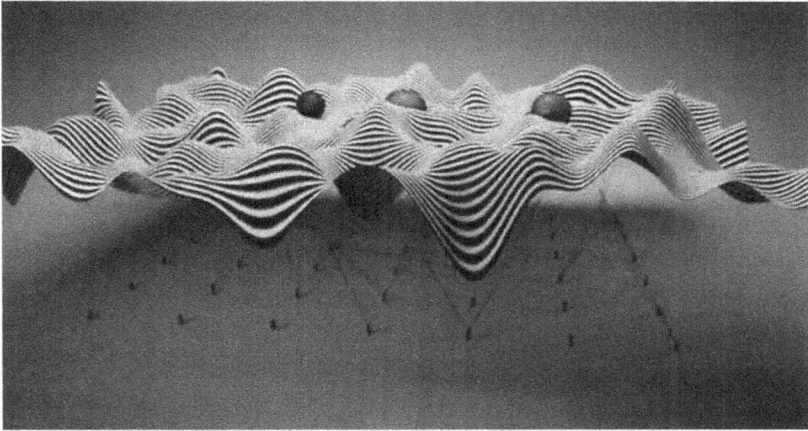

FIGURE 4.9 An animated image of the epigenetic landscape, part of an exhibition of "artists . . . inspired by the exciting world of epigenetic phenomena" (Towler and Harrison 2012).

to help it communicate with the public.[33] Towler created the image for a project on "Epigenetic Landscapes" carried out by Paul Liam Harrison, the artist-in-residence for the EpiGeneSys consortium. "Epigenetics leaves the laboratory and enters the world of art!" the EpiGeneSys website boasts. "The exhibition Hashtag—# Visions of Epigenetics explains basic scientific concepts and presents the works of seven artists, who let themselves be inspired by the exciting world of epigenetic phenomena."

Although it is good to see major, well-funded scientific research projects devote some of their energy—and possibly their economic support, as well—to cross-disciplinary initiatives, notice the disturbing unidirectionality of this relationship: the "exciting phenomena of epigenetics" move out of the laboratory and into the cultural realm, where they inspire artists, whose role it is merely to "let themselves be inspired" and then explain "basic scientific concepts." How different this is from the interactive, mutually transformative uses to which the genre of graphic medicine is already being put in the very medical schools where those aspiring practitioners of epigenetic medicine presumably are receiving their training.

I have been arguing that works of graphic medicine are interventions into medical and health-care fields as much as they are into literature, the arts, and social practices. In fact, because of its capacity to "unflatten" experience, increasing rather than reducing the strings that tie us to our environment, comics as a medium makes us "better able to see these attachments not as constraints but as forces to harness" (Sousanis 2015, 135). Although still unfamiliar in medical school curricula and often an unwieldy entry to the clinical calendar—and at times even weird to those who are used to anatomical illustrations or clinical diagrams—the comics genre known as graphic embryos still promises to join graphic medicine as an alternative model for accessing the broader social and cultural questions posed by the field of epigenetics. The field has its allies, too, as we will see in a later chapter: feminist scientists and theorists of feminist science studies who for some time have been offering their own, reciprocal interweaving of art and science in service of a deeper engagement with epigenetics.

Chapter opening page

CHAPTER 5

The River in the Landscape

The form of a living plant or animal is continuously kept in being
in spite of the fact that material is passing all the time through the
system. The form is not only, in many cases, a complex one, *but the
entity by which it is expressed is more nearly comparable to a river
than to a mass of solid rock.*

—C. H. WADDINGTON, *The Strategy of the Genes* (emphasis added)

How does it come about that all rivers finally reach the sea, in spite
of perhaps initially flowing in a wrong direction, taking roundabout
ways, and generally meandering? There is no such thing as the *sea
as such.* The area at the lowest level, the area where the waters actually collect, is merely *called* the sea! *Provided enough water flows
in the rivers and a field of gravity exists, all rivers must finally end
up at the sea.* The field of gravity corresponds to the dominant and
directing disposition, and water to the work of the entire thought
collective.

—LUDWIK FLECK, *Genesis and Development of a Scientific Fact*

In 1969, the same year C. H. Waddington published *Towards a Theoretical Biology: Sketches: An IUBS Symposium* and *Behind Appearance: A Study
of the Relations between Painting and the Natural Sciences in the Twentieth
Century,* a fellow Scot named Ian McHarg published *Design with Nature,* a
pathbreaking contribution to landscape architecture theory that incorporated the perspectives of art, science, and natural history. Although their

fields were very different, the two men shared several qualities: voracious intellectual curiosity, frustration with the intellectually narrowing force of disciplinarity, and an interest in landscape that connected the embryonic cell to the ecosystem.[1] In this chapter and the next, I explore how the potential that Waddington saw in the epigenetic landscape was taken up by McHarg and the practitioners of landscape theory and design who came after him, who have used it as their intellectual and pragmatic influence. My focus moves from McHarg's scientistic, ecologically informed landscape architecture of the 1960s and 1970s through the resurgence of art and design in the 1990s and the 2000s to the contemporary landscape theory of Elizabeth K. Meyer and the nonlinear, process-based work of landscape theorists and architects Anuradha Mathur and Dilip da Cunha.

A river can be a metaphor for the way science—most broadly construed—can shape social reality through the ongoing, reflexive process of homeorhesis. Fleck (1986 [1929], 54) put it well: "Natural science is the art of shaping a democratic reality and being directed by it—thus being reshaped by it. It is an eternal, synthetic rather than analytic, never-ending labour—eternal because it resembles that of a river that is cutting its own bed." We will come to the river as a metaphor not only for natural science but also for social relations by and by, but we begin with a much more realistic image: John Piper's drawing of a river flowing down a valley between increasingly steep and canyon-like banks. This landscape image enacted Waddington's understanding that the processes of flow and deflection, flux and change are fundamental to every living system. A developing being stays stable *even as it changes*. What precisely were the properties in Piper's drawing of the river that Waddington found so helpful as he thought about development?

> It is a river that flows and yet remains stable in the continual collapse of its banks and the irreversible erosion of the mountains around it. One always swims in the same river; one never sits down on the same bank. The fluvial basin is stable in its flux, and the passage of its chreodes; as a system open to evaporation, rain, and clouds, it always—but stochastically—brings back the same water. What is slowly destroyed is the solid basin. The fluid is stable; the solid which wears away is unstable—Heraclitus and Parmenides were both right. Hence the no-

tion of homeorrhesis. The living system is homeorrhetic. (Serres et. al. 1982, 74)[2]

This watery image comes neither from Piper or Waddington but from Michel Serres. He can help us appreciate what Waddington found in the river, for we can hear the philosopher's debt to the biologist in the neologisms that stud the passage like rocks in a stream. Serres uses words coined by Waddington to designate epigenetic processes. "Chreode" (usually spelled "chreod") from the Greek roots for "it is necessary" and "a route or path," refers to the pathway a cell takes as it develops, moving from totipotency (when it is capable of becoming any kind of tissue) to reach its cell fate (when it has become one specific type of tissue, whether part of an eye or a tooth or a neuron). As Serres describes the river, water makes channels as it courses through the "fluvial basin." These channels, scored into the riverbed by the water continually moving through them, are chreods. They stay the same even as the banks around them crumble because the river itself is not a stable thing but a "system open to evaporation, rain, and clouds." The water in the river joins it, over and over, as part of the water cycle. Yet it comes back "stochastically." Serres's word here is crucial: water becomes river not in a predetermined pattern, but through a process of random distribution. That random, probabilistic, ever changing presence of water is, paradoxically, the river's only constant.

"The fluid is stable," Serres reminds us. "The solid which wears away is unstable. . . . Hence the notion of homeorrhesis." Waddington's coinage, "homeorhesis," is the antonym of homeostasis, a term which refers to the property through which adult organisms are able to stay alive by maintaining their internal systems in a relatively stable state. Waddington proposed that a contrasting process was at work in an embryo as it developed. Homeorhesis, the process by which "the organism stabilizes its different cell lineages while it is still constructing itself," enables an embryo to reach the adult state, when homeostasis comes into play (Gilbert 2014, 633). Through the process of homeorhesis, the embryo maintains a stable rate of flow as it moves along that developmental pathway called a chreod. This may seem paradoxical: how can embryonic tissues simultaneously stay the same and trigger ongoing changes? Yet this is precisely what is required for an embryo to develop.

Here my reading of Serres's treatment of flow, or as he spelled it homeorrhesis, differs from that of Bruce Clarke in *Neocybernetics and Narrative*. For Clarke, who prefers Niklas Luhmann's reading, the problematic nature of Serres's argument is that it does not distinguish between the biological and the thermodynamic systems, or between cycle and flow. Clarke cites as a "hymn to ontological nondifferentiation" Serres's assertion, "Nothing distinguishes me ontologically from a crystal, a plant, an animal, or the order of the world; we are drifting together toward the noise and the black depths of the universe, and our diverse systemic complexions are flowing up the entropic stream, toward the solar origin, itself adrift." In response to Serres, Clarke quips, "I confess that I rather value my differentiation from a crystal, while I am also happy to understand that my animal body is, in fact, a symbiogenetic consortium of bacteria that are already, in their minimal but utterly definitive autopoetic natures, cognitive systems in their own right" (Clarke 2014, 74–75). Because I am viewing this material from the perspective of Waddington's entire oeuvre, however, my emphasis lies not on his investigations of embryonic development and differentiation but, rather, on his late-life attempts to connecting the embryo with the ecosystem. As he explained in his concluding remarks to *Evolution and Consciousness* (Jantsch and Waddington 1976, 244), "Embryos, like ecosystems, are multifactorial. . . . Changes in embryos therefore have to be symbolized by trajectories in multidimensional phase space, in a manner similar to that used by [C. S.] Holling to describe his ecosystems."

The fact that Waddington specifies a phase-space diagram as a representation fit to be applied to ecosystems as well as to embryos (remember, this was the image he acknowledged would really have been the most accurate representation of the epigenetic landscape) suggests to me that Clarke's critique of Serres's position is really a question of scale—both temporal and spatial. Close up, we see the closed cycling of biological systems; from far away (in space and in time) we are more able to watch the thermodynamic flow. In short, the principle of homeorhesis serves me in this chapter as a boundary object. In Piper's drawing, homeorhesis is embodied in the flowing river and its shifting banks; in the environment and ecology more broadly, homeorhesis is a heuristic and methodological affordance, a tool for working with the landscape. Homeorhesis is a quality of the epigenetic landscape that is relevant not only to those who are drawing landscapes, as Piper did, but also to those who are designing them.

This story starts with Scottish architect Ian McHarg, who brought the science of ecology into landscape architecture through a remarkable experiment in interdisciplinary teaching. McHarg was also known for introducing a method of collecting, analyzing, and synthesizing environmental data at multiple scales that would lead to the refinement and installation of Geographic Information Systems (GIS) imaging as a foundational tool in landscape architecture. While he also brought the science of ecology into landscape architecture through a remarkable experiment in interdisciplinary teaching we will discuss presently, my emphasis now is on McHarg's commitment to the tacit knowledge that comes with drawing.[3] By the 1990s, the data-driven mode of practice that had come to dominate landscape architecture during McHarg's career had been succeeded by a reemphasis on the epistemological and methodological importance of drawing as a process that facilitates both the architect's engagement with the environment and the designer's creative processes. As we will see, this reintroduction of drawing offers a return to the landscape as site of *"propulsive life unfolding in time,"* a phrase that reveals the conceptual debt this approach owed to embryology and developmental biology, with significant feminist and environmentalist consequences (Corner 1996b, 81; Wellmann 2017).[4]

The chapter concludes with a consideration of the British landscape architect Charles Jencks's work with the epigenetic landscape. Jencks explicitly and directly incorporates the influence of developmental biology in his built design, creating landforms he refers to as chreods, incorporating images of cell division, and planning a landscape project based on the epigenetic landscape. In contrast, as chapter 6 describes, although they describe themselves as unfamiliar with Waddington, the architects Anuradha Mathur and Dilip da Cunha have devised a set of practices that are distinctly different from Jencks's. We might call theirs a homeorhetic approach to watery landscapes. Like the landscape theorist Elizabeth Meyer, Mathur and da Cunha incorporate a range of primary and secondary sources into their landscape designs. As this visual and conceptual model percolates through an intellectual and aesthetic ecosystem, it has important social and ecological consequences, as we will see.

"Ecology" is a charged and multivalenced term that means different things in different disciplines. The philosopher Lorraine Code approaches ecology as situated feminist epistemology, arguing that by drawing on feminist standpoint theory and attending to the anecdotal, messy, and complex array of human experiences often disregarded in empirical scientific investigations, the field of ecology can produce a more robust form of environmental knowledge (Code 2006). Of course, Code does not explicitly address landscape architecture, a field that has entertained little explicit engagement with feminist theory and practice, reflecting the gendered structures that shaped the field as it consolidated and professionalized and that continue to dominate it (Komara 2000–2001; Meyer 2011).[5] However, from within the profession, the landscape architects Chris Reed and Nina-Marie Lister trace three parallel genealogies of ecology: within the natural sciences, in design thinking and practices, and in the humanities.

Reed and Lister remind us that the term "ecology" originated with Ernst Haeckel, who introduced it in 1866 in his *Generelle Morphologie* as a less restrictive concept than biology. The North American lineage of ecology spreads from investigations by the botanists Frederic Clements and Henry Gleason of the relations among different communities of plants and Robert Park's data-heavy human social ecology of the 1930s to the zoological approaches of Eugene and Howard T. Odum, who published the landmark *Fundamentals of Ecology* in 1953 and the systems-ecology of the Canadian C. S. Holling (1973; see also Park 1936). More recently, Reed and Lister (2014, 3) find a shift in the natural sciences from the positivist data-driven approach to an approach that stresses flexibility and adaptability, while in the humanities they distinguish between writers who affirm the attempt to manage a changing biosphere and those who bemoan such an aspiration as misguided. While they contrast McHarg's reliance on exhaustive mapping with more recent designedly open-ended practices, they place him squarely at the center of ecologically based design thought and practice.

A further discussion at the intersection of philosophy, cybernetics, and new media enables us to expand this view of ecology, illuminating the different flavor it has in the work of some contemporary landscape architects. That conversation differentiates "restricted ecology," the mode of ecology that is characteristic of McHarg's work, from the more recent development of a "general ecology," grounded in the works of Gilbert Simondon, among others (Hörl 2008; see also Clarke and Hansen 2009; Mitchell 2010). This vision of ecology is based in what Simondon calls "transindividualité" or transindividuation (Hörl 2008; Simondon 1989, 248; Stiegler and Rogoff 2010). As a rethinking of the relations between individual and environment, transindividuation installs mediation within ecology, understanding it as a mode of exteriorizing life to remember and transmit it from one generation to the next.[6] Although for Simondon the concept addresses the inter-human relations mediated by technical objects, it can also be understood as an expansion of the notion of epigenetics beyond the individual (or the individual cell) to the transgenerational process of networked transformation, through Bernard Stiegler's concept of epiphylogenesis (Stiegler 2012, 3; Vaccari and Barnet 2009). Here again, the feminist implications—one might even say origins—of this view of ecology within landscape architecture remain to be excavated.

Transindividuation? This initially elusive term came to life for me at a performance by the artist Andreas Greiner and the composer Tyler Friedman at the Import Projects Gallery in Berlin in 2015. As I listened in the dark to their collaborative light and sound performance "Multitudes," an audiovisual installation that incorporated fluorescent algae and two grand pianos, I experienced how relations among media, biology, and ecology could produce connected changes across a number of scales of being. An audience of about thirty young Berliners (and a few older foreign visitors) clustered around two grand pianos. Before the lights went out, we could see that a row of two-gallon water jugs ringed the wall on a wide wooden shelf, while other such jugs rested on the piano wires of the open instrument, at which sat a pianist, and still others lay on the closed top of the other instrument, a player piano. We had been informed that each jug, filled with a gallon (more or less) of water, held innumerable, invisible members of the single-celled species of algae known as *Dinoflagella Pyrocystis Fusiformis*. The lights went out, and as we sat there in the velvety dark, we heard a strange sort of music begin to play. The concert in the darkened room, performed by the organist Hampus Lindwall, was composed by Friedman "based on the growth curve of the algae's replication cycle": "Beginning with total darkness and silence, music and light patterns increase exponentially in complexity, intensity, and frequently, until they reach maximum possible variations. Following this peak, the process ends with complete extinction of sound and illumination, just as the algae's capacity for natural light emission has been exhausted by over-stimulation" (Import Projects 2015).

The music itself was percussive rather than melodic, since heavy water jugs held down the piano wires. But with each piano flourish, which varied from tapping and banging to runs of increasingly urgent ascending thuds, some of the algae jugs lit up greeny-blue. The music continued until, gradually, no more piano trills registered, and all was once again silent and dark.

The first movement of the concert was over, and we took a break in the lighted outer room for wine, beer, and conversation. Soon, a second performance took place. This time there was "a doubling of active elements—algae, score and stimulus." I decided to stay standing this time, and the algae in the jug closest to where I was standing produced starbursts of green in response to the music. I was surprised to feel both proud and protective of these tiny performers. Consider what it means to think of these algae, along with the jugs, the pianos, and the composers and artists, as all having agency. Imagine the program notes for such a concert and you will have grasped transindividuation:

> *Dynoflagella Pyrocystis fusiformis*, the fluorescent algae, has flown in from California, where they were previously with the Pacific Ocean. Composer Tyler Friedman hails from London, where he is interested in "spatial distribution, understood as the fragmentation and dissolution of a singular sound source/point of focus, i.e. the stage, the performer, the instrument, the body of the sound; (extended) duration." . . . Andreas Greiner, once compatriot of the algae in his period as a student at the San Francisco College of Fine Arts, later studied medicine in Budapest and Dresden before becoming Meister Schüler of Icelandic artist Olafur Eliasson, known for his spectacular visual meditations on the Anthropocene.[7]

The "Multitudes" audiovisual installation is intended as a commentary on the process of life, its curators Anja Henckel and Nadim Samma explain: "Playing with biological ebb and flow, emergence and disappearance, both movements amounts [sic] to a requiem for biological life cycles, limited resources and mass extinction" (Import Projects 2015). Perhaps since their description is weighted toward a human perspective, they interpret the piece as a requiem. Yet when approached through the concept of trans-individuation, "Multitudes" is clearly not a requiem but a celebration of the interwoven epigenetics of a general ecology. The performance demonstrates that life energy passes through multiple scales, triggering multiple bright and evanescent moments of individuation, which lapse and reform anew. From the moment the pianist (or his mechanical counterpart, the player piano) strikes the piano keys, the concert produces a rhythmic entrainment of algae and concert-goers. *Dinoflagella Pyrocystis Fusiformis* algae flare with light in a rhythm based on the algae's replication cycle and bring about flares in the audience in response. The concert exemplifies Manuel DeLanda's observation that "multiplicities can become capable of forming a heterogeneous continuum," and that these multiplicities can be "meshed together . . . by extending each singularity into an *infinite series*, and defining these series without the use of metric or quantitative concepts" (DeLanda 2013, loc. 1650).

To speak in terms of Greiner's composition, the relations among composer, algae, and the mathematical principles that define algal development are scaled up in the move from the first movement to the second, rendering human beings, pianos, algae, and sound waves a heterogeneous continuum. "Multitudes" exemplifies how a general ecology can "scale up" epigenetics, bringing to it a transindividual perspective.

IAN MCHARG AND ECOLOGY

While McHarg's perspective for much of his professional life was that of a restricted ecology, as I have said, and Waddington's stretched to a more general ecology, interesting parallels exist among their origins, methods, and chosen metaphors. Both men were deeply connected aesthetically and affectively to the landscape, to the Scotland of their youth and, in Waddington's case, to the endangered Britain of the war years, and they both expressed these feelings in moving essays on their memories of Scotland. Both detoured later in their intellectual development from that initial position, relinquishing that aesthetic, affective mode of relating to the environment for a more reductive, deterministic view of the world.

Just as Waddington the biologist surrounded himself with artists who inspired his thinking, McHarg, a designer, invited scientists as visiting lecturers to his classes in the landscape architecture program at the University of Pennsylvania and incorporated their knowledge into *Design with Nature*, the book he based on those classes. Like Waddington, McHarg held contradictory impulses in tension. He used objective, data-heavy, generalized and comprehensive environmental surveys as part of the process of landscape development, yet he was vehement that specific responsibility and even blame should be assigned for any resulting environmental damage (Margulis et al. 2007, 10). Finally, although he described himself as a simple-minded "crypto-pseudo-quasi-scientist," he expressed his mission as a landscape architect and regional designer in a metaphysical metaphor that resembles Waddington's world egg, if seen through the lens of an argument for transdisciplinary informatics:

> Our job is to reconstitute the region and all its processes again, like putting together Humpty Dumpty . . . This is what modern science is: the egg is shattered, all the fragments lie scattered on the ground. The fragments are called geology and physics and chemistry and hydrology and soil science, plant ecology, animal ecology, molecular biology, and political science. There is no one who can put together again the entire system. Information fragmented is of no use to anybody. What we always need to proceed is really the one whole system, the entire region in question, so for design of sensible human land-use somebody has to put it together again. (McHarg 2007b, 31)

Notice that what McHarg is asking for is a re-creation of the "whole system, the entire region"—an ecosystem view, indeed. And he expresses his concern in an image that is even more unsettling than the returned gaze of the chick newly hatched from Waddington's world egg: Humpty Dumpty shattered. McHarg's Humpty Dumpty metaphor reveals his belief that even if we could gather together "data . . . from many sources [to] describe one whole system, only divided by language and by science," that system is still irreparably fractured (McHarg 2007b, 31). Yet as we will see, McHarg's insistence on using the accumulation of scientific data as the method for putting together that fractured region and all of its processes may have been part of the problem.

Earlier, I suggested that the context within which Waddington did his thinking about development influenced the extent to which he was able to access the range of influences that inspired his creative imagination. In McHarg's case, too, the context within which he was working as a landscape architect and regional designer seems to have shaped him away from an engagement with art and design to satisfy the demands of his private and public clients for data that could be instrumentalized. His approach caught on quickly in the United States, finding a ready audience in the municipal planners and developers who were impressed by its integration of scientific information about natural resources as a platform for their strategies of land development. However, just as with Waddington, there was also another McHarg revealed in his nondisciplinary writings: an alternative self that became legible later, in a new context, only to succumb to disciplinary and professional pressures toward instrumentalism. That alternative McHarg is evident in a posthumously published tribute, *Ian McHarg: Conversations with Students/Dwelling in Nature* (Margulis et al. 2007). In that work, which I discuss later in this chapter, a diverse group of editors, including the biologist Lynn Margulis, recover the broadly ecological, nonlinear, and poetic vision that made McHarg's *Design with Nature* such a celebrated work. McHarg's stamp on the discipline of landscape architecture, however, was his innovative methodology.

FROM *DESIGN WITH NATURE* TO *DWELLING IN NATURE*

The beginning of contemporary uses of ecology in landscape architecture came with McHarg's exposure to the geologists, ecologists, psychologists, and epidemiologists who gave guest lectures in his graduate seminars at the University of Pennsylvania and whose influence he then registered in his celebrated *Design with Nature*. The course topic was "Man and Environment," and he offered as topics for their commentary "the scientific conceptions of matter, life, and man; the views of God, man, and nature in the major philosophies and religions" (Herrington 2010, 2).

As revealed in *Design with Nature*, McHarg's approach to landscape planning is animated by an attention to process and form and an oscillation between development and preservation. The tensions between planning and protection are given vivid voice in his description of the natural environment as a process—or, more precisely, a layered amalgam of many ongoing, dynamic but lawful processes—that provides human beings with both possibilities and constraints. Central to his approach is "one basic proposition . . . that any place is the sum of historical, physical and biological processes, that these are dynamic, that they constitute social values, that each area has an intrinsic suitability for certain land uses and finally, that certain areas lend themselves to multiple coexisting land uses" (McHarg 1969, 104). Because he seeks a system-wide understanding of any region in terms of the phenomena it features and the processes it incorporates, and because he believes that scientific fragmentation has shattered the world egg into a dismal Humpty Dumpty, McHarg turns to science to "make the system whole" (McHarg 2007b, 32). "Our job is to reconstitute the region and all its processes again, like putting together Humpty Dumpty" (31).

He does so by layering photographic transparencies in a model he came to call (somewhat ironically, given the term's potentially gendered and domestic connotations) the "layer cake." The role of this layer-cake model in McHarg's three-step process of regional planning resembles Waddington's use of the epigenetic landscape to think through and represent dynamic developmental processes. Just as Waddington would modify the epigenetic landscape from its 1940 to its 1957 version, the layer-cake model also became less analogue and more digital as the years went on, with results that ultimately led McHarg's methodology to be castigated for claiming an impossible objectivity and objectifying "landscape components as things simply to be mapped and quantified" (Reed and Lister 2014, 13).

As a "practitioner in ecological planning," McHarg explains, it is his role to produce a "creative fit" between the landscape and the client by modifying "both the consumer and the environment" (McHarg 2007b, 26). The design must come about through collaboration between a region and the human beings who propose to use it, with his design help. Thus, for McHarg the task of the landscape architect and regional designer is first to assess the possibilities a particular landscape might offer for human uses, and then to weigh the value of any intervention: how much does it demonstrate "creative fit"? What values does it incorporate? His first step as a landscape architect working on a new project is to call together a range of scientific experts and ask them to assemble a systematic inventory of the region's phenomena, beginning with a bedrock geologist, then a meteorologist, a groundwater hydrologist, a physical geographer, and so on. Of each scientist, he requests "the best description natural science can give us of the region that functions as a single interacting process understood in the context of its long past" to assess the suitability of the site for his landscape design intervention (32).

The second step in McHarg's protocol for planning is to identify and assess the processes in the region: "Everything is always changing. Everything is now and has always been the phenomena, the course of becoming something else. To understand what it is now, one must go through the process of finding out what it has been" (McHarg 2007b, 28). Once again, he calls in the experts: "We now ask them to review the material in terms of process. We ask them to tell us what it is now, in terms of what has been. We ask: what are the dynamics of the process" (28)? In this step he asks the scientists to assess the causal relations between processes. While he acknowledges that this "ecological model" can be descriptive, he usually creates "a graphically represented model like a layer cake, in which transparent maps are systematically overlain on other maps. . . . Each identification of these phenomena I put on a single map. Then I overlay one on top of the other" (28). His description of this layer-cake method reveals the great number of processes whose flow and direction are mapped to provide this model of the "regional 'universe'": "At the bottom is the oldest process, bedrock geology; the next level is surficial geology; the surficial geology can be interpreted in terms of groundwater and surface water hydrology; if possible, the surficial geology is interpreted in terms of physiography; the physiography is interpreted in terms of soils and their variability; the plants in the soil are seen on top; the animals, including man, are then seen on top of the plants; finally, we have the macroclimate, the mesoclimate, and the microclimate" (32).

He may even go beyond those scientific experts, he acknowledges, if a particular site calls for it, including "physical oceanographers, limnologists, and so on" (McHarg 2007b, 32). So to understand and plan for the human landscape that is the ultimate responsibility of a landscape architect, McHarg employs cultural and technical resources to arrive at what he believes is a complete assessment of the processes that constitute it.

At first, McHarg's layer-cake model was distinctly analogue and time-consuming: "All the maps were made as transparencies. The group of relevant factors for each prospective land use was assembled and photographed" (McHarg 2006 [1969], loc. 2792). Different tones of gray had to be incorporated into the map to indicate different kinds of value it represented, and care had to be taken to make sure that when maps were overlaid, the colors did not become muddy and lose their meaning. Once assembled, both the layer-cake model and the matrix associated with it would be presented to a group of stakeholders (the term regains its materiality when we think of the territory to be mapped, staked, and programmed), who would be asked to assess the relative importance of the various values each represented.[8] The projects on which McHarg and his colleagues were engaged in the 1960s and 1970s exemplified that layered approach: they began by assessing the values of processes found in sites that were understood as inherently in flux, including a river basin, the mixed farmland and forested hills outside a rural area, and a city itself. They enabled the designers to give a quantifiable value to things that previously did not have measure. Over the course of McHarg's career, the time-consuming, colorful transparencies that had been the centerpiece of town meetings (and that recall to me Waddington's epigenetic landscape in their graphic, visual, interactive, and nonlinear format) gave way to digitized maps produced by computer scanners "in fewer than ten seconds" (McHarg 2007b, 33). Whether analogue or digital, however, McHarg's layer-cake process produced an ecological model that he believed was "not only . . . a descriptive model, but . . . a predictive model as well. Because the system is an interactive system" (33).

McHarg believed that the process of taking the full inventory of any landscape, in both matrix and layer-cake form, requires assessing the way energy flows among scales—from the microscopic being to the human organism and, finally, the entire ecosystem. In the conclusion to *Design with Nature*, he explains that his ecological inventory is designed to reveal "an ecological value system in which the currency is energy. There is an inventory of matter, life forms, apperceptive powers, roles, fitness, adaptations, symbioses and genetic potential" (McHarg 1969, 197). Note that he is speaking of landscape design in terms closer to thermodynamics than to sociology, psychology, or philosophy. Although it is system-based, McHarg's approach to planning would be critiqued in its later years for excessive reliance on quantifiable knowledge over design intuition. As his layer-cake model extended the practice of mapping developmental trajectories, he hoped it would be not merely thickly descriptive but predictive: "We should be able to say, 'if you affect the system here, there will be repercussions.' We should be able to predict something about the location and nature of these repercussions" (McHarg 2007b, 33). Holding that "ecology offers . . . the science of the relations of organism and the environment, integrative of the sciences, humanities and the arts," McHarg instead claimed that this transdisciplinary field "shows the way for the man who would be *the enzyme of the bio-sphere*" (McHarg 1969, 197; emphasis added).

This metaphorical reference to enzymes, proteins capable of growth and development, requires some parsing. What precisely did McHarg mean? Certainly, this enzyme metaphor expresses his attraction to the science of epidemiology as a model for environmental assessment: "Of all the attempts to make ecological models, perhaps the most useful is that developed by the epidemiologists" (McHarg 2007b, 37). By that interpretation, McHarg's "ecological inventory" resembles the diagnostic protocols and organ systems checklists used by physicians to make sure that they get a full picture of a patient's health (Spirn 2000, 108). Not only does it enable the ecological planner to gather in the full range of information about a region and to present it to the design clients, but it also acts as a prompt for clients to consider issues they might otherwise have overlooked—in particular, the coexisting and competing values embedded in any region.

We could even read McHarg's method as a move to decenter the human, making the environment instead the true patient in need of care. His layer-cake model captures a homeorhetic process of development that is likely to continue, whether or not human beings play a part: "We do not worry here about plants and animals, because our real concern is man. The only truly endangered species is man. . . . If man is eliminated by atomic holocaust or any other man-made device, I think we can be quite sure evolution will start again. The algae will laugh. 'Next time no brains,' they will say, and evolution will proceed into some lovely new expression" (McHarg 2007b, 33).

Yet there is another aspect to McHarg's "Theory of Creative Fitting" that takes no such decentering perspective. This consists of the complex theory McHarg formulated, drawing on ideas from Darwin's *The Origin of Species* and Lawrence Henderson's *The Fitness of the Environment* (1958), that saw the human-environment relation as one of co-constituting adaptation for survival. Central to this perspective is the belief that "fitness" is an evolutionary necessity, casting those individuals and ecosystems who are "unfit" as evolutionary failures. There is a manifest harshness to this normative perspective, as McHarg expresses it: "Our language conforms to this notion of the unfit as the unhealthy, crippled, deformed, although there may well be excellences that overcome this. Beethoven transcended deafness. So unfitness would include not only the broken piano, but also the defaced painting . . . the house in shade or the glaring street, the anarchic city; these are all unfit" (McHarg 1969, 170, quoted in Herrington 2010, 7).

McHarg's "Theory of Creative Fitting" was a curious blend of scientific and religious triumphalism.[9] It was troubling in its disregard of the social issues and oppressive processes underlying the English landscape gardening he valued so highly and deeply flawed by its framing of "ecology as a kind of heavy-footed religion" dominated by the Panglossian belief that the earth evolves along with its biotic and abiotic inhabitants so that "all is for the best in this best of all possible worlds" (Herrington 2010, 15).

Although McHarg continued to believe that "planning and design are a tool of human evolution," the social and economic climate of the 1970s made it more and more difficult for him to move his thoughts from design to built projects. The tradeoff between his ecological interest in mitigating environmental impact and the additional expenditure such design elements would require became more difficult to sell to clients. Thus, Woodlands, the design plan his firm completed for a new town of more than 150,000 people north of Houston, Texas, which specified the "adaptive strategies" that would enable the community design to protect the interacting processes of water flow and drainage and forest growth, was never fully implemented (Spirn 2000, 110).

This difficulty in moving from concept to built projects may have led McHarg to accept his last, deeply flawed commission in 1975.[10] But the muddled combination of denial and zeal that he was now exhibiting may also have explained what went wrong with this final project. The name chosen for the project, Pardisan, was borrowed from the "Old Persian word 'pardis,' from which along with the familiar ancient Greek cognate, it has developed in both Middle Eastern and European languages into the world 'paradise'" (Mandala Collaborative 1975, title page). The client was Iran's Department of Education, and the proposal that the firm of Wallace McHarg Roberts and Todd prepared was a park that combined a cultural history museum, a botanical garden, and a zoo, in the dry and inhospitable region just outside Tehran (Spirn 2000, 110). McHarg's firm planned this desert park as a living example of its layer-cake method: the plan would embody the idea of evolution, with different sections illustrating the processes taking place in different climatological and biological regions. The matrix diagram for the project has category headings that move from Tundra, Coniferous Forest, Deciduous Forest, and Grassland; through Dry Scrub and Woodland, Desert and Semi Desert, and Savannah; to Tropical Forest, with columns beneath each showing the water demanded, the plants and animals and human cultures associated with it, and the "major adaptive strategies" characteristic of each zone. The plan resembles a kind of Aramis in Iran, for as with the failed project in personal rapid transportation that Bruno Latour memorialized in his "scientifiction" of the same name, the Pardisan project was doomed by bad decision making and its planners' failure to accommodate to sudden social changes—in this case, the fall of the shah of Iran in 1979 (Latour 1996; Spirn 2000). The project collapsed, leaving such debts in its wake that McHarg resigned from his firm, without fully understanding the implications of the Pardisan debacle. "When I lost my office," he told Ann Spirn (2000, 112), "ecological planning lost one of its greatest practitioners."

As Spirn describes it, the Pardisan project was ambitious and far-reaching, yet remarkably out of touch with the actual demands of the region for which it was envisioned, whether one considers the water and energy cooling it would have required or the biological and social complexities of the animals and people it proposed to house or attract. As a co-author of the project, Spirn was in a position to know, and her analysis is convincing. However, two other aspects of the park also merit mention—and, indeed, regret: its ambition to decenter the positivist Western worldview by incorporating Iranian mythology and culture and what we might call its holism run wild. The project's declared mandate was to "revers[e] the traditional separatism and reductionism of the sciences" and "reintegrat[e] . . . knowledge for human use" (Mandala Collaborative 1975, 5). Enlisting the help of R. Buckminster Fuller on "epistemological evolution," among other experts, the firm planned to integrate the metaphysical concept of the traditional Persian garden into "a selective representation of the world in microcosm," a "planar simulation of the world with Iran in the center" (30). Despite its aspiration to "represent the world in microcosm [and to] present the world from the vantage point of Iran," however, Pardisan was Janus-faced in its adaptation of the Iranian perspective (9). Promising to provide opportunities to enhance human health through studying the epidemiology of adaptation, as well as the challenges of environmental toxicity and geopathology, Pardisan also planned to provide surveillance, via time-lapse photographic images of Iran at a resolution to one acre, "revealing the country through changing seasons," that would enable forecasting, monitoring, and interpretation. The images would be obtained by the U.S. ERT satellite that, the proposal notes, passes over Iran every eighteen days: "Its four spectral images may be employed in the form of photographs with a discrimination of 100 acres. These images can also be represented with a resolution of 1 acre. . . . It is also recommended that this technique be used for planning inventories, for forecasting, monitoring and, most valuable, for interpretation, research and education" (10). Initially envisioned as a kind of Biosphere-cum-Disneyland, the Pardisan project proposal reveals McHarg as an anxious Humpty Dumpty, his attempt at a transcultural, trans- species paradise verging on a panopticon.

What if McHarg had practiced ecological planning differently? Would his ecological inventory have taken a different form if he had been able to resist the involuntary internalization of disciplinary hierarchies that led him to elevate the scientific ones over the aesthetic? If he had been able to resist pressures ranging from the new allure of satellite technology to the demands of geopolitics? As with Waddington's autobiographical introduction to the second Serbelloni Symposium, McHarg began *Design with Nature* with an autobiographical essay that suggests what the alternative path might have been. Elsewhere I have argued that such modes of thought and self-expression enable the expression of embodied and affective knowledge beyond the bounds of scientific rationality precisely because they are un-disciplinary and even undisciplined (Squier 2004). In their fictive, even poetic tones, such personal asides represent an act of saying otherwise. Here I want to take that argument a step further: I now understand those modes of thought as alternative and deeply powerful methodologies for doing scientific work.

We can see this methodology at work in the autobiographical essay "City and Countryside" with which McHarg opens *Design with Nature.* He begins by explaining that "since evidence tends to be too cold I feel it more honest and revealing to speak first of those adventures which have left their mark and instigated this search." A topographical imagination suffuses this essay. We meet McHarg as a child poised between the road downhill to Glasgow, with its tenements, factories, and shipyards, and the route uphill into the countryside, to "Craigallion loch and the firepot where hikers and climbers met, the Devil's Pulpit and the Pots of Gartness where the salmon leapt, as far from my home as Glasgow was" (McHarg 1969, loc. 290–308). Much like Waddington's "autobiographical note," this explicitly sets out the tensions from which McHarg's work springs and the theme that dominates it. Class pressures channel him in his trajectory to becoming a landscape architect, as he is propelled away from the grim streets of industrial Glasgow toward the beautiful highlands. "I have found that it has been my instincts that have directed my paths and that my reason is employed after the fact, to explain where I find myself. Hindsight discerns a common theme, astonishingly consistent" (loc. 330).

In three vivid moments, each of them with a subtext resonant of the economic context for landscape design, McHarg's narrative illuminates what would be his work's consistent theme. The first is the devastating discovery that his beloved Peel Glen had been destroyed by the spreading city of Glasgow. He describes this with the force of a prose poem:

> Lark and curlew, grouse and thrush had gone, the caged canary and the budgerigar their mere replacements. No more fox and badger, squirrel and stoat, weasel and hedge-hog but now only cat and dog, rats and mice, lice and fleas. The trout and the minnow, newts and tadpoles, caddis and dragonfly are replaced by the goldfish alone; the glory of beech, pine, and larch, the rowan and laburnum, the fields of poppies and buttercups, the suffusion of the bluebell woods are irreplaceable—in the gardens are some desultory lobelia and alyssum and sad, brave privet shoots. The burn is buried and water now is the gutter trickle and spit. (McHarg 2006 [1969], loc. 383)

What he mourns, with luscious specificity, is not merely the land lost to a housing development but the variety of animals and plants that once lived there, now replaced by sad, domesticated, even verminous animals, insects, and plants. Even the wonderful water that filled the burn has become just another form of human refuse, the abject "gutter trickle and spit." The reduction of this landscape from complexity, variety, and health to austerity, uniformity, and disease is detailed through an inventory of the different species and natural processes lost to development, a vision of the human being as inextricably shaped by (and shaping) his environment. The framework here is one of general rather than restrictive ecology, and the intensely particular picture it paints is dramatically different from the generalized inventories accumulated through his layer-cake and matrix methods of planning for development, even if the latter were assembled with a weather eye to environmental effects.

The second experience that shaped McHarg as a landscape architect was developing pulmonary tuberculosis, which required that he stay for "six sweating months" in the Southfield Colony for Consumptives outside Edinburgh. Here, too, the environment seems not merely a container for his body but a space that shapes his soul and restricts his sense of possibilities. "Ceilings are important to a prostrate patient," he explains. "These were of Italian plaster work, deeply configured, and in their recesses were dark spider webs with collections of flies. The entertainment of the place was to watch the blue tits fly into the room and gorge on these insects. The spirit of the place was acrid. . . . [The] sun never shown, the food was tepid and tasteless, there was little laughter and less hope" (McHarg 1969, 396). Only when he escapes that sickroom to the Swiss Alps overlooking Lac Léman does he return to health, an emotional awakening that he describes as "strong stuff—such an experience engraves the belief that sun and sea, orchards in bloom, mountains and snow, fields of flowers, speak to the spirit as well as the flesh, or at least they do to me" (424). Here, too, the quality of the environment as an active agent is powerfully productive of his emotional and physical health.

The last experience McHarg mentions as key to the common theme of his work as a landscape architect and regional designer is a summertime visit to New York City. As a Scot, he tells us, he finds the heat not only unfamiliar but actually intolerable. Exhausted, wrung out, and thirsty after a day spent touring celebrated landmarks of modernist architecture, from the Museum of Modern Art and the United Nations to Lever House, he enters a brownstone renovated by Philip Johnson and stands with his hosts, looking out through a window onto a garden:

> This was dominated by a pool with three stepping stones, a small fountain, a single aralia tree and on the white painted brick walls, a tendril of ivy. We stood on a narrow terrace beside the pool, savoring the silence, then discovering below it the small noises of the trickling foundation, drips and splashes, the rustle of the delicate aralia leaves, seeing the reticulated patterns in the pool, the dappled light. Here were these selfsame precious things, but consciously selected and ar-

rayed, sun and shade, trees and water, the small sounds under silence. What enormous power was exerted by these few elements in this tiny space. (McHarg 1969, 437)

Water, light, movement, sound: these vibrant agents in the environment have a power that McHarg feels moved to express directly. But ironically, as palpable as are the emergent and affective resonances of this passage, so is the privileged economic context that enables them. Such class calculations are left behind in the essay's concluding paragraphs, which adopt a general ecological perspective to call for "not only a better view of man and nature, but a working method by which the least of us can ensure that the product of his works is not more despoliation":

This book is a personal testament to the power and importance of sun, moon, and stars, the changing seasons, seedtime and harvest, clouds, rain and rivers, the oceans and the forests, the creatures and the herbs. They are with us now, co-tenants of the phenomenal universe, participating in that timeless yearning that is evolution, vivid expression of time past, essential partners in survival and with us now involved in the creation of the future. . . . Let us then abandon the simplicity of separation and give unity its due. . . . [Man] must become the steward of the biosphere. To do this he must design with nature. (McHarg 1969, 464)

Reading this passage again, on Earth Day 2016, I find myself wishing that McHarg's legacy had been able to incorporate the vision of interspecies co-tenancy glimpsed here, rather than—as in the Pardisan project—committing to the scientific explanation of nature as the definitive goal.

Such a shift in understanding requires a different perspective on science itself, of course, that conceives of it less as a prescriptive project than a collaborative choreography. This view survives in *Ian McHarg: Conversations with Students/Dwelling in Nature*, a slim volume published in 2007, six years after McHarg's death. The volume allows us to imagine what it might have meant had McHarg been able to design *with* nature in the terms he sets forth here. It provides a record in miniature of McHarg's interdisciplinary ambition and genre-busting daring, qualities that were muted over time by the zeal for scientific data and control that until very recently have been his dominant legacy in landscape architecture. A lively interdisciplinary gathering between book covers, the edited volume demonstrates the openness to genres beyond the conventional disciplinary essay that recalls McHarg's accomplishment in the courses that inspired *Design with Nature*.

The collection—edited by Lynn Margulis, originator of the influential theory of symbiogenesis; the landscape architect James Corner; and the forestry and habitat biologist Brian Hawthorne—features contributions by a professor of design, an environmental historian, two biologists, a science writer, an ecologist, and a landscape architect. Indeed, the volume stretches beyond the life sciences to incorporate literature, including publishing McHarg's poem "Pond Water" (1980) for the first time. Poets and scientists, from Emily Dickinson to Maxine Kumin and Waddington, with his "Bubble Blastula," have long written about the teeming watery world glimpsed through a microscope. The Dutch scientist Antonie van Leeuwenhoek looked through his microscope and discovered the motility of human sperm, as Kumin reminds us. But what distinguishes McHarg's poem is its tight focus on the small creature that he finds under the microscope: not human sperm but the tiny freshwater rotifer.

Rotifers play an important role in keeping fresh water clean by digesting decaying organic matter, which may be one of the reasons for McHarg's interest in these small beings. But this impulse to write about them in a poem suggests that he was moved, here as in "City and Countryside" to use the power of language as well as of science to achieve ecological understanding:

POND WATER
To test
a new microscope
a drop of pond water,
A teeming world,
Creatures shooting across the slide,
Some passing through others.
A rotifer with a propeller
In the middle of its head,
metasynchronized cilia.
Improbably,
as are they all

living out lives
lasting minutes
Are we so different?
(Margulis et al. 2007, 77)

I suggest that a connection exists not only between rotifer and the "bubble blastula," but also, on a larger scale, between McHarg's interest in development as a mode of environmental intervention carried out by landscape architects and Waddington's interest in development as a biological process that incorporates the rhythm of transgenerational and environmental influences. The speaker marvels at the creatures' improbably brief life span, and then, with a sudden shift of conceptual scale, wonders, "Are we so different?" While the metaphysical speculations are familiar, McHarg's poem speaks to the science geek in many of us, as well.[11]

Although it is doubtful that McHarg knew this, the processes his poem captures extend beyond philosophical speculation to scientific discovery, expanding the implications of epigenetics from human development to the entangled development of different species in an ecosystem. In the decades since McHarg wrote his poem, the rotifer has been used as a model animal in experiments demonstrating how environmental influences can activate advantageous developmental changes, a process known as predator-induced polyphenism. Scott Gilbert describes this microscopic process as a dramatic story of lifesaving epigenetic influence:

> Imagine a species whose larvae are frequently confronted by a particular predator in their pond. . . . One could then imagine an individual that could recognize soluble molecules secreted by that predator and could use those molecules to activate the development of structures that would make this individual less palatable to the predatory. . . . Several rotifer species will alter their morphology when they develop in pond water in which their predators were cultured. . . . The predatory rotifer *Asplanchna* releases a soluble compound that induces the eggs of a prey rotifer species, *Keratella slacki*, to develop into individuals with slightly larger bodies and anterior spines 130% longer than they otherwise would be, making the prey more difficult to eat. (Gilbert 2014, 667)

We do not need to stay at the granular level of these research findings to be impressed with how McHarg's thinking in his poem loops together epigenetics and ecology with aesthetics and literature, or to regret that he lost the ability to do this in his design process. Yet while his holistic ambitions clashed with his restrictive ecological vision, I suggest that the problem was also, and again, an environmental one—a problem with his environment as a landscape architect. I mean that the tensions in McHarg's work reflect general tensions in the field itself: between those who take science as the reigning authority and those who grant equal agency to designers, as well as between those who view the mission of the profession as prediction and control and those who see it as design and collaboration.

"THE MAKING OF MARKS AND THE SEEING OF POSSIBILITIES"

In the years since McHarg dominated the field, landscape architecture has swung back to an appreciation of how art and design can also serve an ecologically informed perspective that is more broadly construed. Corner, one of the best known landscape architects practicing in the ecological tradition first established by McHarg, has reframed what such a perspective can include. Observing in 1995 that ecology has a problematic ambiguity that may be responsible for the familiar question that dogs landscape architecture—"Is it art or science?"—Corner traces this ambiguity to the two models of ecology that dominate the field: the "conservationist/resourcist [model], which espouses the view that further ecological information and knowledge will enable progressive kinds of management and control of ecosystems," and the alternative restorative and protective model, which stresses healing and restoring endangered ecosystems (Corner 1996b, 90). We can hear echoes of McHarg in both models, of course, but what Corner adds is the view of art and design not merely as illustration of an already accomplished mode of management but as methodologies for an emergent creative practice. Arguing that the field's overemphasis on scientific data has prevented it from emerging as "a representational and productive art . . . a cultural project," Corner offers a number of important strategies to bring the aesthetic and design-based aspects of landscape architecture back to the forefront (85–86).

Although it is tempting to discuss all of Corner's strategies for bringing design back into the field of landscape architecture, I will restrict myself to mentioning only three of them: his redefinition of drawing practice as central to the profession; his theoretical commitment to a practice posture directly opposed to the epistemological certainty at the heart of the conservationist/resourcist view; and his embrace of an aesthetic indebted not to the tradition of the English landscape garden but to Chinese and Japanese landscape painting.

For Corner, drawing is not the illustrative postscript to a design process but its essential creative core. Landscape architecture has made a major mistake in its understanding of drawing, he argues: it has either viewed it as the highly valued artistic product of the design process (which leaves the actual work of landscape architecture design to the side, unassessed, even unnoticed) or as an aesthetically valueless instrumental adjunct to a primarily technical process (which ignores the creative vision the drawing may express). In contrast, Corner advocates thinking of the act of drawing as both prior to and outside the encounter with the landscape. In his view, the act of drawing enables the architect to *think through on paper* his approach to a place. Using the practices of projection, notation, and representation, the landscape architect can express analogically the architect's growing plan, projected into the picture plane; record the temporal and spatial layers of experience produced by his or her encounter with the landscape; frame the landscape before the design work has been performed; and even, if necessary, transform the way the space may be seen by society so the work can be done. Corner locates the very power of drawing for landscape architecture in that paradoxical doubleness: "Landscape architectural drawing gains its potency precisely from its directness of application to the landscape, on the one hand, *and* its disengaged, abstract qualities on the other" (Corner 1992, 264).

The practice of landscape architecture should aim not at exhaustive knowledge but at the ability to dwell in uncertainty, Corner argues. Drawing makes possible a crucial experience of imaginative renewal. Exploratory in its very process, drawing can give rise to "bewilderment, wonder, and indetermination" (Corner 1996b, 100). For Corner, such attitudes, born in the relationship to something beyond the artistic self, are essential for landscape architects. As exemplified by Surrealists and expounded by Henri Bergson, Gaston Bachelard, and Maurice Merleau-Ponty, these attitudes enable perspective necessary for what he calls "a more animate appropriation of ecology in landscape architecture" (104). Thus, Corner proposes that "a truly ecological landscape architecture might be less about the construction of finished and complete works, and more about the design of 'processes,' 'strategies,' 'agencies,' and 'scaffoldings'—catalytic frameworks that might enable a diversity of relationships to create, emerge, network, interconnect, and differentiate" (102).

The final aspect of Corner's argument for bringing drawing back into landscape architecture moves beyond the Eurocentric aesthetic frame emphasized by McHarg's valorization of the English landscape garden and even by Corner's references to Surrealist art and phenomenological philosophy. In addition to thinking of drawing as a medium that translates an idea into "a visual/spatial corporeality embodied in the built fabric of the landscape," he argues that drawing must also be understand as "the *locus* of reconciliation between construal and construction, or between the symbolic and instrumental representations" (Corner 1992, 265). What Corner means by this is exciting: he argues that the very act of drawing combines both "the making of marks and the 'seeing' of possibilities." Notice the similarity between this process-based understanding of drawing in the work of a landscape theorist and the function of drawing as an act of self-materialization in cartooning explored in chapter 4.

As an example of this collaboration between artist and environment, Corner offers Chinese and Japanese landscape paintings of the fourteenth and fifteenth centuries that employ the improvisational strategy known as "flung ink." He writes, "Ink is first thrown onto the canvas in an energetically random manner to form a visual field. The painter then improvises through immediate response to the thrown image and begins to construct a landscape through the workings of the brush" (Corner 1992, 265). Two sections of a long scroll by the Japanese Zen Buddhist artist Sesshū, *Splashed Ink Landscape* (1495), particularly exemplify this process for Corner, in which the process of flung-ink painting opens up a "synesthetic 'field,' a metaphorically suggestive realm that prompts an imaginative seeing." As he explains in a caption to *Water and Mountain*, by Sesshū, "Process, work, duration, accident, flux: while landscape is the subject, equally so are the painter's own spontaneity and involvement with the brush" (Corner 1992, 267).

Another scholar has taken the argument even further, arguing that *Splashed Ink Landscape* exemplifies "painting degree zero, in which ink painting appears to signify nothing more than its own genealogy, and its own mastery" (Lippit 2012, 50). Corner's appropriation of the methodology of flung-ink landscapes for his design practice gives us landscape architecture degree zero: a mode of conceiving and executing an ecologically inspired project for landscape design that emerges from, and echoes, the metaphoric, stochastic vitality of the epigenetic landscape extended to a molar scale. And like the Japanese tradition of painting that unites ink flung randomly on the page with the painter's improvisational brushwork, Corner views the process of drawing as a metaphorical spur to creation in landscape design. The drawn line connects the designer, the materials of drawing (charcoal, pencil, paper), and the landscape, bringing all three together in a homeorhetic process of construal, construction, and creative change. Thus, a general ecological vision emerges from the new way of seeing made possible by this art process.[12] Corner chooses the term "deixis," borrowed from the art historian Norman Bryson, to describe the way that symbolic or speculative and representational or demonstrative drawings work together to create the end product of landscape architecture. "The deictic drawing," he points out, "records and traces its own evolution, and refers back to an entire corpus of prior thoughts, ideas, and associations" (Corner 1992, 273). Corner selects three works of a landscape architect in training, Anuradha Mathur, then a graduate student in his design studio, to exemplify how such deictic drawings work. As he explains in a comment on Mathur's "Planometric Collage of an Ecological Garden" of 1990, "While the predominant mode of presentation can be understood as a plan, hidden within the drawing are other views and images. The aim of the drawing is to try to embody as much of the symbolic ideas and intentions as it does the instrumental" (273).

This chapter began with thinking about the homeorhetic properties of water (sometimes rain, sometimes clouds, sometimes ground seepage) aided by the philosophy of Michel Serres. I then tested the gendered boundaries of the term "ecology," as taken up in landscape architecture. In the next chapter, I return to this question and explore how deictic drawings, along with painting, silk screening, singing, walking, and traveling by boat, work to enable collaboration with the changing properties of water in the landscape architectural process of Mathur and da Cunha. I close now by looking at the more direct and less improvisational mode of incorporating Waddington's model of the epigenetic landscape into built design: Charles Jencks's landforms.

Jencks is a British architect, landscape designer, and sculptor, as well as the author of numerous scholarly books on postmodernism and improvisatory design, building styles, and urban architecture. These beautiful coffee-table books, which include *The Garden of Cosmic Speculation* (2003) and *The Universe in the Landscape* (2011), lushly document Jencks's own landscape architecture designs, as well as the visually rich environments he has created for himself and his clients. Jencks's house in London's Holland Park neighborhood intrigued and charmed me when I visited him in March 2015, from its stunning ammonite-inspired spiral staircase ("Solar Staircase") to its garden path ending at a door with mirrored panels that bears the slogan "The Future" and, below it, "Is behind You."[13] Jencks is a generous and engaging man, and our conversation revealed his passionate resistance to reductionist thinking, attraction to science at its undisciplined margins, and appreciation of an artistic and analogue approach to form that he shares with Corner. As Jencks explained to me, "Everybody is digitally driven today. Everything we do. And the digitals can out beat analogues but we clump along. . . . No question that everybody bemoans the fact that no one can draw anymore in architecture. No one can."

Originally from Baltimore, Jencks studied English literature as an undergraduate, then moved to London to study architectural history. He now maintains homes in London and Scotland. In our conversation, Jencks described his intellectual influences as a landscape architect as incorporating "Waddington through Arthur Koestler and through Lancelot Law Whyte [a Scottish engineer and industrialist], and . . . the intellectual wing of the non-specialist biological thinkers or philosophers" (Jencks interview 2015, 3). Models of all kinds have long interested him not as representations of scientific truth, but as practices that enable scientific exploration. As he explained, "We architects and landscape people use models to understand and to think about in a way that allows us to interact physically—very crudely—but importantly I think, because you don't have to be as accurate" (3).

The interest in models, metaphysics, and science is in evidence at Bonnington House, a Jacobean manor house outside Edinburgh where Jencks's works are on exhibit in a one hundred-acre sculpture garden called Jupiter Artland. There, one can visit "Metaphysical Landscapes," an exhibit of what Jencks calls "small-scale landforms and studio works," and "Cells of Life," a landscape that involves "eight landforms and a connecting causeway [that] surround four lakes and a flat parterre for sculpture exhibits."[14] Jencks describes the "Metaphysical Landscapes" sculpture installation as "content-driven and hybrid design," an amalgam of art, design, text, and landscape: "My miniature landforms are where I try out mixing media—of writing, signs, and the art of nature itself, particularly strange and beautiful rocks. We have a parity with nature and that metaphysical balancing-act leads to a style and artform."[15] "Cells of Life" draws on embryology, particularly the process of mitosis, the event in nucleated cells in which cell division begins and an organism starts to develop from a single-cell zygote. To one visitor to the Edinburgh landscape, this piece of landscape design revealed "an uncanny relationship to division of membranes and nuclei" in its division into two landforms (Azzarello 2013).[16]

Yet what, precisely, is uncanny in Jencks's work? Not the use of scientific imagery and concepts, which was already incorporated by many of the artists Waddington discussed in *Behind Appearance*. And while the sheer number of scientific concepts Jencks referred to during our conversation (from genetic determinism and epigenetics to chaos theory, strange attractors, and the multiverse) was at times difficult to harmonize, that experience was more dizzying than uncanny. Listening to the descriptions of his projects, I was impressed by the bubbling enthusiasm of his bricolage. But as we continued to talk, Jencks made a comment that gave me pause. He explained that these big concepts were difficult to realize because he was required to weigh the site plans incorporating those challenging, exciting visions against soberly realistic assessments of the professional climate: "We don't occupy all the positions of tenure and power in the world, and we live in a mechanistic atomistic reductionist world. Remember that in the end Bill Gates, who supports all this, and that world, wants results tomorrow. They want results, and the money will always force the game for an answer tomorrow."[17]

That comment clarified for me what is uncanny about Jencks's work: the disconnect between the deeply felt inspiration that clearly motivates it and the economic demands and material impact of his mode of landscape creation. I mean by that not only the fact that his very large installations require the expense of petroleum and chemicals in their creation and maintenance (land-moving equipment to make the mounds, large lawnmowers to preserve the meaningful line of the installations, and fertilizer to keep those beautiful swirls of lawn green), or that a Jacobean manor house holds the art foundation where the works are displayed, but also that in most cases (though not all) a large amount of money must change hands before the projects can come into being. Although we are no longer in Jacobean England, it is clear that the status quo economicus of the society, however unjust or hierarchical it is, must remain stable if projects such as the ones at Bonnington House are to achieve completion. Both Woodlands and Pardisan, McHarg's unsuccessful late projects, come to mind.

Let me be clear: My critique here is not of Jencks himself but of the way such landscape architecture functions not as an interaction with the land that changes both parties in the exchange (think back to Waddington's discomfort at the returned gaze of the little chick hatched from the world egg) but as an instrumental imposition. Rather than the landscape also shaping the planner, this process of landscape architecture enforces human agency on passive terrain. Jencks's attachment to the landscape draws him in two conflicting directions: to intellectually themed abstraction and to embodied and affective intervention. For the former, there is his massive "Garden of Cosmic Speculation," at Jencks's home in Scotland, Portrack House:

> Forty major areas, gardens, bridges, landforms, sculptures, terraces, fences, and artificial works. Covering thirty acres in the Borders area of Scotland, the garden uses nature to celebrate nature, both intellectually and through the senses, including the sense of humor. A water cascade of steps recounts the story of the universe, a terrace shows the distortion of space and time caused by a black hole, a "Quark Walk" takes the visitor on a journey to the smallest building blocks of matter, and a series of landforms and lakes recall fractal geometry.[18]

Despite the obvious pleasure Jencks has taken in making the land express his aesthetic and intellectual fascinations, the terrain is clearly recipient rather than generator of the rather abstract intellectual speculations it now hosts, and the effect is one of hyper-individuation rather than transindividuation.

For the latter, in contrast, we have the embodied and affective exchange that characterizes Maggie's Centers, the garden-enclosed buildings initially co-created by Jencks and his wife, the architect Maggie Keswick Jencks, when she was diagnosed with cancer. Now a Scottish registered charity, the drop-in centers for people with cancer combine buildings designed by prominent architects and gardens that aim to provide an "intense connection to nature," according to the British garden designer Dan Pearson. Advocating an approach that links the environment, health, and the powerful affect of hope, the centers feature gardens designed not only to raise the spirits of cancer patients, but also to enable emotional and physiological healing.[19]

Landscape architecture as a profession exhibits an ongoing tension among the methodologies of data collection and scientific assessment; processual, improvisational drawing; and the goals of control and collaboration. In the next chapter, I explore what happens when landscape architecture takes on the very riverine landscape with which this chapter opened, approaching it not as a field for content-driven metaphorical design but, rather, as an ecology, a physical site where land and water meet and seem to call out for development.

CHAPTER 6

Designing Rivers

In the mid-1940s, three thousand German and Italian prisoners of war, selected for their knowledge of engineering and construction, worked with members of the U.S. Army Corps of Engineers (USACE) to create "the world's most ambitious working model" of a river (Cheramie 2011; Mathur and da Cunha 2001, 14). After clearing and grading forty acres of land just outside Jackson, Mississippi, they installed many thousand feet of drainpipes and then poured, shaped, and cut more than a million yards of earth into a bas relief of the Mississippi River basin (Mathur and da Cunha 2001). Once cast in concrete, this one hundred square foot segmented model was placed on pilings that could be raised or lowered to control the elevation. The purpose of the model was to monitor the river's flow in flood times, to demonstrate visually to the general public just how well the corps could control the river, and to help USACE test and demonstrate projects it might build in the future to increase its capacity to control where the river would flow.[1]

Conceived of and overseen by USACE's district engineer, Major Eugene Reybold, the model functioned from the late 1940s until the Mississippi flood of 1973. Heralded as making possible the first panoramic rather than segmented view of the river, the concrete cast modeled the Mississippi and its tributaries as "a system, a network of continuous forces that creates unique but interconnected conditions" (Cheramie 2011). To some, the Mississippi River Basin Model (MRBM) was an attempt to respond to the threat a rising river posed, and it proved its value during flood season in April 1952 when it enabled engineers to identify and evacuate areas most likely to be inundated. One could argue that it had still another goal in mind: the provision of a sense of security, however illusory. Visitors from up and down the river could climb its observation tower, and when they did, they saw a landscape scaled to make the vast river system accessible to a human viewer: "A vertical foot on the model equals one hundred feet of the Mississippi landscape, a horizontal step on the model is a mile, and 5.4 minutes in its working is a day in the real world" (Mathur and da Cunha 2001, 14). The MRBM gave viewers and engineers alike the comforting sense that, since it was possible to map the river's edges, center, and trajectories, it might also be possible to control them.

Despite that dream, however, the MRBM was ultimately abandoned (Cheramie 2011). It was just too costly, in dollars, labor power, and political will, to maintain as an ongoing process. Once the German prisoners of war were repatriated, it became difficult to find a steady stream of laborers to operate the model. Maintaining funding was also difficult, because the money was allocated not by state but by district, and it had to be "equitably divided among 15 districts in proportion to their river frontage" (Cheramie 2011, 7). Since the model was based on a fixed-stream bed design, with the river channels and flood plain areas built of concrete, its very fixity constrained and distorted the system it was attempting to represent. Pre-cut concrete channels left no space for the flowing system that is the Mississippi and its tributaries, an ever changing territory ribboned by rivers and constantly carved anew into channels and islands. Neither the banks with their artificial foliage created of "accordion-folded metal screen" nor the river, with its "parallelepieds," brass plugs introduced into the concrete "to create drag in the water flow and simulate scouring," captured the vital unpredictability of a real river to find new paths, spread, twine, and diverge across the landscape. (In fact, the MRBM's designers cut costs in the end and never incorporated the mouth of the river or its delta.) In the end, weeds moved into the concrete channels where they had never been expected to grow; the land grew in around it; and this once working topographic model was replaced with computer modeling. Data now did what cast concrete could never do.[2]

The creators of the MRBM were engineers preoccupied with prediction and control. The model had a successor, however, one conceived as a response not to an ecosystem in need of stabilization but to an ecosystem

FIGURE 6.1 René Thom's hydraulic model of the epigenetic landscape, like the "hydraulic model of reproduction" shown here, employs a representational strategy earlier used by embryologists: the clay model. The caption points out that these models were "kindly built" by M. Marcel Froissart, the French physicist (Waddington 1970, 99).

understood to be homeorhetic. I am referring to the mathematician René Thom's hydraulic model of the epigenetic landscape (figure 6.1).

We met Thom in chapter 1 as one of the participants in the Serbelloni Symposia. A French mathematician and part of the Bourbaki group (a mathematical publishing collective named for Nicolas Bourbaki and founded by the topological mathematician Henri Cartan), Thom was also an interdisciplinary thinker "by temperament," argues the applied linguist and biosemiotician Donald Favareau, who recalls a colleague's observation that "Thom was forced to invent catastrophe theory in order to provide himself with a canvas large enough to accommodate the diversity of his interests" (Favareau 2010, 347). Thom's interests would expand to include not only mathematics and biology but also animal and human communication and even different societal structures. But at the third Serbelloni Symposium, as part of an argument that topological models of morphological development should extend beyond the biological realm, he presented a "hydraulic model" specifically based on Wadding-ton's epigenetic landscape (Waddington 1970, 99).

As Thom described the model, it expressed the principle that "the phe-nomenon of 'life' itself is best thought of as a processual and 'lifelong epigenesis'" (quoted in Favareau 2010, 350).[3] Acknowledging that such a model is not "amenable to experimental control," he argued that its value lies not as a scientific theory but, rather, as a method: "This method does not lead to specific techniques, but strictly speaking, to *an art of mod-els*." Such an art will always be essential, he said, in situations in which, despite our desire to achieve control, we find ourselves rather in "diffi-culties, or contradictions . . . or when we feel overwhelmed by the mass of empirical data without a clear notion of the problems at hand" (Thom 1970, 115). Modeling is important under situations of affective stress, he asserted, because models allow us to stay afloat in a flood of quantitative information. As if in direct response to the data-heavy project of the MRBM, Thom asserted that what we need is "a qualitative understanding of the process studied": "Our dynamical schemes . . . which remind us of the old Heraclitean ideas . . . provide us with a very powerful tool to reconstruct the dynamical origin of any morphological process. They will help us, I hope, to a better understanding of the structure of many phenomena of animate and inanimate nature, and also, I believe, of our own structure" (115).

Thom's reminder that we need to take the model as a tool for under-standing rather than as the thing itself might have been intended for the Louisiana landscape architect Kristi Dykema Cheramie. Walking through the MRBM in 2011, "consumed with the immediacy of the experience," she was struck "by the disconnect that can occur when a model becomes a substitute for 'the real thing'" (Cheramie 2011, 14). When she asked her-self what had gone wrong with the MRBM, she was able to extract three lessons: Materials matter (the materials with which the model was cre-ated were inadequately complex to simulate the real thing); scale matters (the distortion in scale required to make a manageable model may have led policy makers to draw false conclusions and create bad policy); and scope matters (the plan was flawed from the beginning by the decision not to incorporate any of the river below Baton Rouge).

Cheramie's conclusions are convincing, but she misses the most impor-tant lesson of all: mind-set matters. That is to say, the way one frames the river will constrain the success of the model one builds. The makers of the

MRBM viewed the river as being in need of control, and their model was constrained by the limitations of that approach. They needed a different mind-set.

Before turning to several landscape architects whose theory and practice demonstrate that new mind-set, I want to pause to examine a powerful formulation of it available online, the Australian cartoonist and journalist Sam Wallman's web comic, "So Below: A Comic about Land." In chapter 4, we considered how the comics medium can give us one more way to deal with the world in which we live, thus enhancing the "explanatory pluralism" that is a "positive virtue in itself" (Keller 2003, loc. 3159). Wallman's black-and-white scroll comic contributes to that explanatory pluralism by examining the most basic of concepts: the notion that human beings can own land and all that follows from that principle: "the notions of private property, land as commodity, national borders, exclusion zones, and the everyday resistance to these ideas from Indigenous people, refugees, the homeless, grassroots communities, animals and plants."[4]

Rather than a stable linear narrative, the comic is presented as a loop of images accompanied with text configured on the screen as if it were poetry. Presented as moving pixels rather than as stable text, the comic scrolls down without borders as if performing its very argument. After surveying the extractive relations that human beings in all of their dreadful variety historically have had with the land, it concludes bleakly, "I think, / That we would be well served / To calmly remind ourselves occasionally / That we all live in outer space. / & that the way the dust has settled on the present day . . . / Is nothing more / Than concrete absurdity, / Informed by the past. / . . . concrete absurdity." Spiraling back to its beginning, the comic concludes with our embodied need for space that now seems to reveal "a quiet desperate mania." Yet exploiting the pun on which it will soon settle, it offers in passing hope that lies in the agency of the land itself. "Concrete can crack. I seen it B4." That new mind-set is the innovation of the landscape architects and theorists to whom I now turn.

I mentioned in chapter 5 that landscape architecture has engaged little with feminist theory and practice. However, as long ago as 1997, in *Ecological Design and Planning*, the American landscape architect Elizabeth Meyer articulated her vision of a new, expanded "theory for the built landscape." Meyer explicitly defines herself as a feminist theorist in her essay, and she draws not only on landscape architecture, but also on feminist philosophy, postmodern philosophy, feminist literary criticism, feminist art history, and feminist anthropology, to formulate what she frames as five principles for a new way to approach the land. Meyer's feminist principles for landscape theory resemble in many ways the approach of the landscape architects to whom I devote the bulk of this chapter: Anuradha Mathur and Dilip da Cunha. But in order to clarify precisely how Mathur and da Cunha expand on and complicate Meyer's feminist landscape theory and practice, I excerpt her argument here, with apologies for the length of this passage:

1. Interpretations of built works and treatises should be based on primary experiences that are mediated through the knowledge of historical situations. This primary experience has two forms—visiting a site; and studying historical plans, maps, treatises, journals, letters, photographs, and the like.

2. We should be suspect of generalizations that "transcend the boundaries of culture and region." Instead, theoretical work should be contingent, particular, and situated. Grounding in the immediate, the particular, and the circumstantial . . . is an essential characteristic of landscape architectural design and theory. Landscape theory must rely on the specific, not the general; and like situational and feminist criticism landscape architectural design and theory must be based on observation, on what is known through experience, on the immediate and the sensory—what is known by all the senses, not only the eye. Thus landscape architectural theory is situational; it is explicitly historical, contingent, pragmatic, and ad hoc. It is not about idealist or absolute universals. It finds meaning, form, and structure in the site as it is. The landscape does not sit silent awaiting the arrival of an architectural subject. The site—the land—speaks prior to the act of design.

3. We should be skeptical of discourses that assign a gender affiliation to the landscape—implicitly or explicitly. The implicit affiliations are manifest as "female"—the "other" who is seen but not heard. . . . The explicit affiliations are manifest as "feminine"—that which is irrational, wild, chaotic, emotional, natural.[5]

4. While the deconstruction of the discourses that relegate landscape to a silent female or irrational feminine role in modernism is neces-

sary, it is not enough. We need to reconstruct the unheard languages of the modern landscape as a means to reinvigorate contemporary design practice. The work of a feminist design critic is reconstructive, not destructive. This reconstruction assumes a multilayered fabric that weaves together threads from primary sources and documents written by landscape architects and about landscape architecture with the concurrent history of ecological ideas, cultural and historical geography, design and planning criticism, and site interpretation.

5. Finally, landscape architectural history has been, for the most part, a masculine discourse focusing primarily on the works of great landscape architects—mostly men. . . . This historiography must be enhanced and challenged, for it denies the conditions of practice, conceptualization, and experience. This challenge exposes landscape history as a fiction that has been written through a particular lens or sensibility that has ideological implications. . . . Let us propose that landscape architectural history and theory should be about the cultural, geomorphological, and ecological history of the preexisting site as well as the history of the design project and its designer. . . . The intersection of geometry and geomorphology, of past site and present project, requires a dialogue between the site as a speaking figure and the designer's markings on that site. Landscape design is not about monologues. . . . This double-voiced discourse is predicated on a systems aesthetic, not an object aesthetic; it is about the relationship between things, not the things alone. (Meyer 2011)

Meyer powerfully proposes these principles as an alternative to binary constructions of the landscape and our relations to it and to each other; in contrast, she advocates for a "spatial continuum of response that unites."

Joan Scott has written about feminism as a "restless critical operation, as a movement of desire" detached "from its origins in Enlightenment teleologies" (2004, 19). I hope we can apply that perspective as we return to the Mississippi River in the discussion that follows, keeping Meyer's manifesto in mind, in terms of where it resonates with the work of Mathur and da Cunha that I discuss and where it presents interesting contrasts due to the frame and limits of such central concepts as primary experience (what is it and for whom?), race (where is it?), and history (whose history?).

"THE MISSISSIPPI CARRIES A DESIGN AGENCY"

Anuradha Mathur, a landscape architect, and Dilip da Cunha, an architect and city planner, have spent more than a decade working with watery landscapes, from their first major U.S. project, *Mississippi Floods: Designing a Shifting Landscape* (2001), and including their pair of projects in India, *Deccan Traverses: The Making of Bangalore's Terrain* (2006) and *Soak: Mumbai in an Estuary* (2009), and, finally, their ongoing participation in "Structures of Coastal Resilience" (Mathur et al. 2014b), the multisite project supported by the Rockefeller Foundation. Their ambitious, theoretically compelling, narratively rich and visually beautiful installations, art books, and works of landscape theory merit a full-length study, for they offer a more fully elaborated perspective than I can sketch out here—in particular, their recent, dazzlingly interdisciplinary collection of theoretical essays, *Design in the Terrain of Water*, which introduces the concept of the river as the product of "the drawn line, a product of a visual literacy rather than a natural feature of the earth surface, an extraordinary work of art before it is a taken for granted object of science" (Mathur and da Cunha 2014a, 1).[6] They describe their distinctive approach to landscape as "art before science; art that questions the 'things' and their visualization that 'experts,' such as engineers, ecologists, and planners, often take for granted" (Mathur and da Cunha 2009, xii). Technically in the lineage of Ian McHarg, with whom they share the use of both the layer-cake and matrix methods of scaled ecological inventory in their ecologically informed approach to design practice, Mathur and da Cunha bring a very different mind-set to the riverine landscape. They draw their inspiration from Anaximander of Miletus, who first identified what is now called the "hydrological cycle," but which he saw " 'as a river with a circular course' that 'continues indefinitely. ' " It is this "extended time of water," they argue, "The grand cycle of movement of water from ocean to atmosphere to continent and back to ocean [that] is the essential mechanism that allows organisms—including humans—to emerge, to develop, and to live on earth" (Mathur et al. 2014a, quoting Leopold 199, 2). "The idea of the river comes with the drawn line and there are many instances in which actually it reveals itself as a very problematic idea because when rivers flood they're basically crossing the line that we have drawn. And so we ask, "Why do we see it as problematic?" Because it is our own making. So the notion, actually, of linear entities actually moves from source to destination.[7]

Homeorhesis happens not only in the river (of Piper, of Serres), but also above and below it, because water also exists as evaporation, clouds, and rain. I learned this from Mathur and da Cunha during an interview I conducted with them both on March 5, 2015, as water fell (as snow) outside their window in Philadelphia: "Rain has always been put in the service of rivers. But rain can also be another moment in time wherein you don't get a linear flow but you get a nonlinear flow system that actually moves in depths across air, water, land, sea, earth, sea. You know, it sort of dissolves those distinctions in a very rich depth and a multiplicity of flows and you get a much more complex system that we believe could be an alternative ground of design" (da Cunha 2015).[8] By framing the drawn line that defines a river as more important than the water that forms it, they unleash the design potential of water as multiplicity.

McHarg was trained in Cambridge, Massachusetts, on the banks of the Charles River, no longer a "stinking tidal estuary" since the landscape architect Charles Eliot built a dam across its mouth, and now "the man-made Charles River Basin," according to the Charles River Watershed Association.[9] In contrast, Mathur and da Cunha began their training in regions of India where rivers were historically unstable beings, drying up in the summer like the Sabarmati in Ahmedabad or surging across flood plains like the Yamuna River in New Delhi.[10] While their later work takes the concept of an estuary as an inspiration, and the notion of man-made river margins as part of the problem, in their first major project, *Mississippi Floods*, the mindset they were prepared to challenge was the notion of "a landscape of flood."

When Mathur and da Cunha turned to the Mississippi in 1996, they were responding to devastating images of the great Mississippi flood of 1993 and the hardship it caused: "breaking levees, moving houses, desperate sandbagging, and anguished faces" (Mathur and da Cunha 2001, xi). But they were also motivated by their sense that responses to this flooding were polarized between those who viewed it as a natural disaster and those who saw it as the result of human interference with the river's flow. And they were disturbed with what they found as the dominant view of the Mississippi River itself as "an object, an object that can be controlled or released, an object that can be subject to categories." Instead, they thought of the river as something shifting, elusive, and propulsive; they decided to join the Mississippi in its travels "to try to learn from it what other processes coexisted with the great river." Rather than accumulating and presenting scientific information about the region under study (like McHarg's systematic method), Mathur and da Cunha thought of their "journey through the Mississippi mud" as a rhythmic mirroring of the river itself, "truly the inhabiting of a shifting terrain" (xii).

If McHarg's accomplishment was to bring to landscape design the holistic vision of an ecosystem combined with the predictive statistical power of epidemiology, the achievement of Mathur and da Cunha is to reconceive landscape design as an immersive process that enrolls experts and amateurs, locals and outsiders, flora and fauna, and, most important, the landscape itself. There are significant similarities between their sense of themselves as landscape designers and the sense proposed by Elizabeth Meyer in "The Expanded Field of Landscape Architecture" (1997). They view themselves as being engaged in a design process that collaborates with, rather than attempts to control, all of the other processes within a landscape. The attention to chance, situatedness, positionality, complexity, multiplicity, and the power of juxtapositions to catalyze new thinking are all crucial aspects of their work. As they explained to me, when they are beginning a project, they usually choose four or five "starting points of engagement," and their approach to those starting points emphasizes serendipity, chance, and the unleashing of possibilities through cultivating difference. To understand what I mean by this, and what its implications are for how their landscape architecture projects unfold, let us pause to consider how they described their design process to me in an interview:

> We work one against the other and we associate, by chance, certain things with certain things, but in order to play out their differences more than actually their similarities. So we are always actually on a relative ground rather than on an absolute ground so we keep playing that relativism for as far as we can go. That is one way of actually making chance, I would say, intrinsic to the way in which we investigate, and sometimes the chance leads to a good thing; sometimes it's a dead end, and we start again. But that is very much part of the process. . . .

When we actually look at issues of certainty and uncertainty, which is often what comes up in conversations today, we say that the opposite of certainty is not uncertainty. The opposite of certainty is possibility.

So there are two ways in which we actually frame possibility, if you look at the screen prints of the photo works. One of the things that we say often is that we take the world that has been simplified today—say, into a map—it's been simplified in a map that water is here, land is here, . . . and we embed it in a complexity, whether it is in a screen print or in a multiplicity of layers or . . . a multiplicity of juxtapositions, we immerse it in a complexity so that when we pull it back up, it comes up differently. And we never know quite how it comes up differently. That is chance. To some extent, we play with that element of chance and design, and then we can take it further. So when we are playing with five different starting points, we allow each one of them to come up to resurface from the complexity of a screen print or the complexity of a folder work. We allow them to surface differently. That is one side of the way in which we allow chance. We give it a kind of an opening.

The other thing is that when we actually finish a project for exhibition, as we did in Norfolk, or as we have done in Bombay, we are always at pains to say that we don't have the singular solution to anything, and we don't begin with a problem that is to be solved. What we're doing is actually opening up the imagination, and what we say is that we have just opened up one thing. If you had to follow a process of investigation as we have, you may come up with something else. And so, chance actually then is peculiarly our own. And we allow other people to then say— of course very few people take on the challenge—but it is another way of actually living in a process rather than actually outside the process. And so, we're making process actually very much the act of design.[11]

In our interview, Mathur and da Cunha contrasted their process-based model to what they called "the McHarg world": "McHarg actually still remained outside the process, framing the process, etcetera . . . But his design methodology actually never allowed him to be immersed in the process."

In *Mississippi Floods*, their vision of landscape design is apparent in the Mississippi-based design process itself. It is lyrical rather than programmatic, tentative and open to mistakes and blind alleys, and engaged with a variety of sites they describe with the chunky complexity and vital variety of a "Latour litany": "catfish farms and factories, gas sta-tions, revetment fields, experimental forms, churches, motels, bed and breakfasts, hunting clubs, crossings, even the Delta Regional Medical Center, where one of us was admitted with dehydration" (Bogost 2012; Mathur and da Cunha 2001, xiii). Seeking alternative perspectives on the river in pictures, histories, folktales, and songs and drawing inspi-ration from John McPhee's book *Control of Nature*, they also talked to people in all walks of life, consulting sources both socially and profession-ally central, as well as marginal, even workaday sources: lawyers, blues singers, an agricultural researcher, a pilot, and a riverboat captain: "Each day brought new revelations, introduced us to new ways of seeing things, ways that constantly changed the course of our travels." They emphasized multiplicity even in the media they used. As they "stumbled upon sites and followed leads" they documented their findings in a variety of media: "map-prints (screen prints in which we layered and erased information), photo-transects (photomontages in which we brought together the maps and horizons of our journey), historical maps, line drawings, and paintings" (Mathur and da Cunha 2001, xii).

The whole complex set of vital sources comes together in their volume *Mississippi Floods*, whose title (with its verb/noun ambiguity) memorably expresses their doubled vision of the Mississippi as a place and an agential subject. By approaching the Mississippi as a working rather than merely a scenic landscape, they situate their project as part of the larger move to "reveal representations, particularly maps, as powerful ideological, co-lonial, cultural, and even fictional instruments in the service of power" (Mathur and da Cunha 2001, 8). They ask not only what those authorized maps and charts reveal but also, more important, what they obscure. They start with the MRBM, which they see as an example of a longer process: "the dematerialization of the Mississippi for the purposes of its control [that] reaches back to well before the advent of computers . . . to the early maps and surveys that sought to define its boundaries, to contain its ho-rizon, first in the service of empire and then in the service of the em-pirical science of river hydraulics" (15). Then, working to bring back those aspects of the Mississippi that official maps and charts leave out, they explore the river's design agency as it carves the landscape into meanders, basin, flows, banks, and beds. By following those four modes of design and tracking the Mississippi in space and time, they reveal the ontological instability of what professionals take to be its basic categories—"river, settlement, water, soil." They intend the book that emerges as the record of this journey to seed a public project "directed not toward resolving the problem of flood but toward keeping the possibility of reimagining the Lower Mississippi alive" (153). In its affirmation of diversity, serendipity, chance, change, and growth, Mathur and da Cunha's *Mississippi Floods* is a powerful embodiment of the river at the center of the epigenetic landscape. Less than a decade later, the devastation of Hurricane Katrina led the feminist philosopher Nancy Tuana to revise an essay of her own to broaden its exploration of the connection among female embodiment, ecology, and interdisciplinarity—or, as she put it, "the materiality of the social and the agency of the natural":

> Seeing through the eye of Katrina transformed an essay that was focused on women's embodiment to the embodiment of levees, hurricanes, and swamps as well as the embodiment of the women and men of New Orleans. . . . I would argue that at the core of this essay is the centrality of an interactionist ontology as the lens through which we must be feminists and do our feminism. . . . . it is only through erasing the dichotomies that have been erected between the natural and the social, including the natural and the social sciences as well as the arts and the humanities, that we can craft a material feminism. . . . Interactionism not only allows but compels us to speak of the biological aspects of phenomena without importing the mistaken notion that this biological component exists somehow independent of, or prior to, cultures and environments. (Tuana 2008, 190, 209–10)

Although they do not explicitly label their work as feminist, Mathur and da Cunha's work and vision share much with that of Tuana and the feminist landscape architect Elizabeth Meyer, beginning with the interactionist—or we might even say, intra-actionist—view of the landscape (Barad 2007). When they extend that transdisciplinary, interactionist ontology to India, to explore the historical and cultural origins that shape its riverine landscape, we find them expanding the field of landscape architecture still further, putting pressure on the meanings it attributes to the concepts of history and development.

Flooding was the concern when Mathur and da Cunha were called to Mumbai one year after the damaging monsoon rains of July 2005. Again they were impressed by how a conceptual frame could restrict possible approaches to landscape design. As they describe in *Soak: Mumbai in an Estuary* (2009), public and government administrators alike thought of Mumbai through the "lens of flood," focusing on ways to keep water in its proper place. Here again, the problem, as they saw it, was mind-set. The region had been conceptualized through the lens of flooding, as they found when they explored the history of the area. During British rule, regional maps were drawn only in the dry season, and planning was based on the fixed city those maps revealed. Administrative decisions "not only assumed the course of flows, but also enforced them with property lines, walls, embankments, and other constructions" (Mathur and da Cunha 2009, 72). Yet when the monsoon hit in 2005, the flooding did not occur as people expected it to. Rather than river-borne floods, "flood waters rose from a saturated ground even as they fell in sheets from a 15 km thick ground."

The difference between expectations formed by maps and this unanticipated reality challenged Mathur and da Cunha to engage in a process of design that Arjun Appadurai and Carol Breckenridge have called "wet theory": "a way of building explanations and models which accommodates flux, flow and other boundary-blurring phenomena at the core of theory rather than at its reluctant boundaries. . . . theory that recognized its own uncertain footing, that is humble before the ruthless tyranny of context, and that is always ready to negotiate with the facts that sweep up against its shores or rain down on it from the heavens" (2009, ix). The term is well chosen for its resonances with "weak theory," "theory that comes unstuck from its own line of thought to follow the objects it encounters, or becomes undone by its attention to things that don't just add up but take on a life of their own as problems for thought" as part of an investigation of the forms of living (Stewart 2008, 72).[12] In dividing the "sciences of measurement planning and control from those of interpretation, representation and poetics," Mathur and da Cunha choose options often conceptualized as "wet" or "weak" in contrast with the dry rigor of positivist science. Wet or weak theory begins with the act of "unthinking" the assumptions that have habitually structured our approach to place, particularly "what we mean when we contrast the categories of land and sea" (Mathur and da Cunha 2009, x). They question the conceptual frame they encountered, even as they acknowledge its seeming urgency, given the flood of the previous July. They ask: is it really helpful to think in terms of flooding and flood prevention? Where did that kind of thinking begin? What have been the historical effects of that approach to Mumbai? What alternative and more productive models might now be found for approaching the region as a design field? Through their multimodal practice of site research, "walking, drawing, and photographing what maps cannot depict, books cannot explain, and *Slumdog Millionaire* can only speed through," they assemble a social, cultural, and political history that testifies to the fact that "the war against the monsoon" extends back to the days of British colonial administrators (xi).

Let us recap their method. Arriving at a Mumbai structured by the framework of flood and defense against the monsoon, they "unthink" that framework, proposing instead that Mumbai be thought of as an estuary, "a terrain that operates more as a filter between land and sea than a line between them" (Mathur and da Cunha 2009, 7). (The biological resonances here, to permeable membranes, seem to me worth noting.) They move beyond the habitual adherence to one of the core principles of landscape architecture, McHarg's layer-cake method. Instead, they introduce in *Soak*, a new model for representation and visualization: the section. The term is both conceptual and methodological. Conceptually, "section" structures the book's design into three parts. The first part is a survey of how the history of mapmaking—in particular, the act of drawing a line separating land from water—has led to the "war against the monsoon" that has structured the city. The second is an alternative investigation of the territory based not on geographical space but on time. To convey their findings here, they chart the relations between land and sea, juxtaposing sections drawn in sequence to learn the temporal rhythms that structure them over time. (We found resonances of this when we considered the rhythmic seriality enabling the development of experimental embryology, as well as the sequential juxtaposition central to comics.) And the third is a proposal for the territory, consisting of twelve different "initiations," each of which "works to resolve the problem of flood not by enforcing lines, but by transforming Mumbai into a place that absorbs the monsoon and sea, a place that accommodates soak" (8).

Methodologically, a "section" or "photosection" is "a photographic sequence that cuts through time and space with rhythm and purpose," a model for visualization and representation that offers an alternative to the map, with its dominating, all-seeing God's- or surveyor's-eye view (Mathur and da Cunha 2009, 192). The section offers another way to conceptualize the region. Operating on the vertical rather than the horizontal axis, it draws attention to depth rather than spread and to "intersecting continuums rather than finite adjacencies" (7). Working between the fixed map and the fluid section, Mathur and da Cunha demonstrate that Mumbai's relationship with water should be solved "not by flood-control measures, but by making a place that is absorbent and resilient" (ix).

Documenting the "build-up to war against the monsoon" on a number of fronts, they reveal its pervasive effects through charts, screen prints, maps, photographs, historical documents, and graphic images. One section of their investigation can convey the complexity of their practice. Project 4, Sion Fort, is an assessment of one of the many "creek forts" built by the English colonial administrators to control access to part of the estuary crucial for imperial trade. Mathur and da Cunha provide not only the historical frame, but also the layer-cake assessment that defines all of the processes essential to this specific site within the Mumbai estuary in a double-page spread. The image provides multiple temporal and spatial scales to capture the site as a living entity and uses sections to incorporate the flora and fauna of the site (people, monsoon waters, fish), the processes (gathering, celebrating, filtering, treating, cultivating), and the outlook (gathering, viewing) (Mathur and da Cunha 2009, 118–19).

Mathur and da Cunha return, but with a reparative difference, to a practice deeply rooted in colonial history: the exhibition of "maps, plans and other visualisations" that the city planner Sir Patrick Geddes compiled and took on tours through several Indian cities in the 1910s and 1920s to generate a shared perspective and frame for the government planning to come (Mathur and da Cunha 2009, 4). Finally, in accordance with their view that design should provide an incentive to imagination rather than a finished product, they submit their design plan in a public exposition rather than a white paper to the Indian government. To the "problem-solving and largely reactionary mode of governance that traps places following a disaster" they oppose their "critique of flood and the possibilities of its alternative" (x).

Central to this vision, as to so much of the work as landscape architects, is the act of reconceptualizing Mumbai's relations to the water. As they explained to me in our interview, Mathur and da Cunha view Indian history as founded on the erroneous assumption that Bombay [Mumbai] is an island and thus that "all the events in history are actually told on the basis of island thinking":

> So when we introduce . . . a different landscape, we do it in history as much as we do it for the future, because we say that history is written for the future; it's not written for the past. . . . With a different ground, we see different events taking place. So if Bombay is an estuary . . . as opposed to Bombay the island, the estuary actually presents a completely different challenge. You cannot assume that people arrived at an island. They arrived in a much more complex environment that they made into an island. So the island is an act of creation just as much as the river is actually made by Alexander as an act of simplifying the landscape because he was colonizing a land. He was not colonizing people; he was colonizing the land. So you realize that there is a landscape prior to history.[13]

This discussion helps to clarify the particular relation that they take to the postcolonial project. While they acknowledge the power of the subaltern school's critique of Indian history, they find that it fails to question one fundamental assumption. "Unable to imagine another ground," they argue, "they use the same maps to be critical maps" (Mathur and de Cunha interview, 2015). Instead, they maintain that the very act of mapping—of drawing those lines defining water and land—has conceptually created the very landscape on which that Indian history has been built. The position recalls the critique of mapping expressed in "So Below: A Comic about Land" (figure 6.2), in a sequence (borderless; the comic cannot necessarily be described as having panels or gutters, a topic in its own right) depicting how "Things / got particularly / buck-wild / when the practise [sic] of / Cartesian mapping / emerged—/ grids of straight lines / laid over the round earth / in order to solidify property ownership."

By viewing Bombay as an estuary, "a dynamic threshold between salt waters and monsoon waters; a third coast after the edge with the sea on

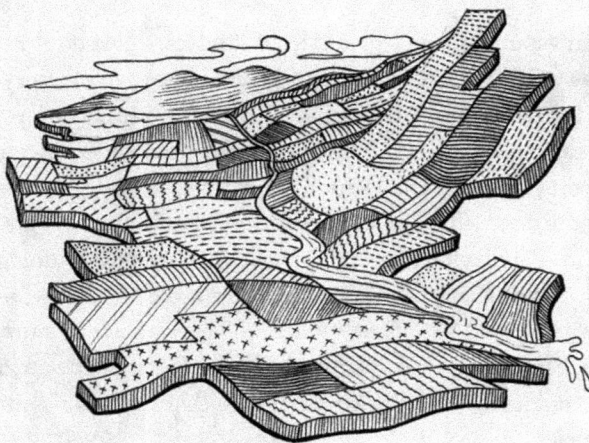

FIGURE 6.2 Sam Wallman's "So Below: A Comic about Land" incorporates sequence and cartographic images to challenge the concept of land ownership.

the west and harbor on the east," Mathur and da Cunha (Mathur and da Cunha 2009, 190) show us how to move beyond the simplicity of island thinking to appreciate the political, climatological, biological, hydraulic, and human processes that together have shaped, and continue to shape, the city of Bombay. Stirring testimony, rich historical and cultural archive, and collaborative art installation, *Soak* is intended, they explain, as both "a catalogue of an artistic endeavor that challenges conventional visualisations of Mumbai as well as an extended arm of an activist agenda that encourages the expectation of new possibilities for 'another' Mumbai" (xi).[14]

The final landscape architecture project of Mathur and da Cunha that I discuss is "Turning the Frontier," their contribution to the multi-site Structures of Coastal Resilience project, which is supported by the Rockefeller Foundation and, according to its website, is "studying and proposing resilient designs for urban environments on the North Atlantic coast of the United States."[15] Under the architect Guy Nordenson of Princeton University, this larger project draws together elements of landscape planning that, as we have seen, are frequently in tension or held at some distance from one another (Spirn 2000). It incorporates a design team, a GIS modeling team, and a team engaged in engineering and science research, with focuses on storm-surge modeling, sea level rise, and coastal climate change. In its first phase, the project included four teams: the Harvard Graduate School of Design; City College of New York, Spitzer School of Architecture; Princeton University School of Architecture; and the University of Pennsylvania School of Design. There were also four design sites, chosen for their likelihood to experience significant coastal flooding: Narragansett Bay, Rhode Island; Jamaica Bay, New York; Atlantic City, New Jersey; and the area around Norfolk and Hampton Roads, Virginia. The project has now been whittled down to three groups, one of which is Mathur and da Cunha's "Turning the Frontier."[16]

The problematic that this massive multisite project addresses is coastal flooding arising from climate change or global warming—or, as it is called on-site in Maryland and Virginia (since both of those terms are off-limits in those locations) "recurrent coastal flooding." Once again, rather than superimposing a method of controlling that flooding, Mathur and da Cunha have decided, as in Mississippi and Mumbai, to learn from the water. They have adapted the same strategy with their design team that they used on those earlier projects: "When we went with our first field trip with our whole crew, we just sort of dispersed them in different directions and said, 'Look at this. Oh, you gave us structure but this is how we are going to travel.' And when we found the histories we were looking for, we were able to start" (Mathur interview 2015). The histories they were looking for, once again, were not the human histories structured by and structuring the idea of a coastline that needed to be defended, but the earlier histories engraved by water itself, in the streambeds of the multiple rivulets and rivers of the tidewater country. As they explain, "The inspiration for a coast that is cumulative rather than continuous comes from the tidewater country of Virginia itself" (Mathur et al. 2014c, 48). "The University of Pennsylvania team's project, then, is to turn the coast of Norfolk and its environs so that land meets the sea not across a "front" but, instead, through a number of discrete fingers of high ground. Each finger is a unique gradient or a unique gathering of gradients between land and sea, working to structure a coast that is more fractured, cumulative, and diverse than it is continuous, linear, and absolute" (48).

"Turning the Frontier" offers an opportunity to see the homeorhetic nature of Mathur and da Cunha's landscape design practices in action. What I mean by this is that they incorporate key strategies from earlier projects while also adapting their approach as those strategies are deflected and redirecting their project in response to the particular demands of this collaborative venture. What are those key strategies? As in Mumbai and on the Mississippi, they begin with "unthinking" the assumption that coastal flooding is a problem to be solved. Instead, they seek design strategies based on a vision of the coast as porous, dynamic, and reciprocal, a continually changing set of relations between the sea and the land. As in their Mumbai project, they propose a set of "presentations, publications and exhibitions" addressing not only the people with the power to plan, but also those who are "in need of empowerment" (Mathur et al. 2014c, xxvi).

Their proposal demonstrates a rhetorical shift reflecting the fact that, as they admitted in their interview with me, "we were a little more aware in Norfolk." Perhaps that is why in "Turning the Frontier" they discuss the design principle that grounds their project by using terminology drawn from their project partner, USACE, the very agency they criticized in *Mississippi Flood*. Borrowing the language of the USACE report, they explain that they propose to create "nature based features" that "mimic characteristics of natural features but are created by human design, engineering, and construction to provide specific services such as coastal risk reduction" (Mathur et al. 2014d). These "nature based features" are what they call their Fingers of High Ground, the structures that incorporate and extend a key feature of the tidal river: its ability to "point, extend, bend, reach, fold, grow, nudge and retract," to "gather and work a number of gradients—dynamic and shifting reciprocities and trajectories between land and sea, . . . gradients [that] are operational, material, temporal, spatial and ecological" (Mathur and da Cunha 2014c, 149). Operating "at multiple scales simultaneously," these Fingers of High Ground will enable the passage of water, plants, and animals back and forth between the sea and the inland terrain, enabling processes of slow adaptation to changing water levels rather than making the doomed attempt to create an impermeable coastline through "barriers, gates, and pumps" (Mathur et al. 2014d, 26).

Ludwik Fleck (2012 [1935]) stressed the powerful chreodic nature of thought communities, which, because they possess the dominant and directing disposition akin to a "field of gravity," can ensure that rivers do indeed flow toward what they will then agree to call "the sea." In its rhetorical tactics, Mathur and da Cunha's "Turning the Frontier" exhibits the ability to change when deflected while continuing to develop a design strategy. Their approach as landscape architects and regional designers returns us to Fleck's analysis of the river, that served as the epigraph to chapter 5. Although they share with Fleck the appreciation that rivers can flow "in a wrong direction," take "roundabout ways," and "generally meander," they challenge his belief in an inevitable outcome, determined by gravity or discursive *gravitas*. Drawing widely on different communities of thought, experience, and practice, Mathur and da Cunha resist the normative mapping of a river, the disciplinary norming that can curtail the variety of processes they use in their design strategies, and the geographic and political overdetermination that can interrupt our ability to respond to a site on its own terms. Instead, in their works in the United

States and India, and in their landscape theory, they call our attention to the possibility of another sort of development: one not produced through the imposition of control from outside, whether that is USACE or the British Colonial administration, but that springs from the design potential inherent in the river itself.[17]

CHAPTER 7

"A Complex System of Interactions"

Art Laboratory Berlin as an Epigenetic Landscape

On a hot and sunny day, July 20, 2014, I rode my newly purchased second-hand bicycle from Prenzlauer Berg (in the former East Berlin) to Wedding (in the former West Berlin) to a gallery that I had heard hosted interesting exhibitions of bioArt. The field was not a new one to me. I had seen "Victimless Leather," the installation of a tiny tissue-culture leather jacket, in 2008 at the Museum of Modern Art's show "Design and the Elastic Mind," by Oron Catts and Ionat Zurr; the installation had made news by overgrowing its culture chamber and having to be euthanized. I had also participated from afar as a member of the advisory board of the SymbioticA Biological Arts research laboratory at the University of Western Australia in Perth; enjoyed the performative presentations of bioArt by Adam Zaretsky at the Society for Literature, Science, and the Arts conferences of 2007 and 2010; and learned much from Robert Mitchell's discussion of how bioArt operates within an "ecology of relationships" and appropriates "the logic of the fold," bringing together research institutions, the public, and corporations in a new mode of interaction that shocks us out of our usual relation to objects (Mitchell 2010). I particularly appreciated Mitchell's useful mapping of the three eras of vitalist bioArt: concern with plants and heredity; concern with recombinant DNA and innovations in the creation of life; and concern with bioterrorism.

Berlin had introduced me to a new way of thinking about bioArt, however. After the concert by Andreas Greiner and Tyler Friedman at the Import Projects Gallery, where two grand pianos, some imported fluorescent green algae, and the organist Hampus Lindwall collaborated to create a multispecies musical piece based on the algae's replication cycle, I wondered whether bioArt, like landscape architecture and graphic medicine, might be another area where the epigenetic landscape functioned as a methodological tool, a prompt for thinking, and even an alternative mode of practice.[1] I followed that hunch when I bicycled to Art Laboratory Berlin (ALB) on that hot July day. What I found, to my pleasure, was a community testing the possibilities of the extended epigenetic landscape: material, metaphorical, and methodological.[2]

Let us pause a bit to examine the third version of C. H. Waddington's scientific model. Unlike the river and the schematic embryo on the hill that preceded it, this image of the epigenetic landscape explicitly factored in the choreographed interactions of genes and environment viewed as part of a developmental system.[3] When it appeared in Waddington's *The Strategy of the Genes*, this third version of the epigenetic landscape was captioned: "The complex system of interactions underlying the epigenetic landscape" (Waddington 1957, loc. 710). The image reveals the converging forces through which the process of canalization constrains the direction of development, making alternative paths less and less likely. It also reveals how changes at the individual scale of development are connected to those on the evolutionary scale. The meshed and interconnected guy wires create strains on specific points, producing a contoured topography, as the biochemical effects of gene expression constrain the array of all possible developmental choices (figure 7.1). That contoured landscape canalizes the development of the embryo. As Waddington (loc. 710; emphasis added) explained, "The pegs in the ground represent genes; the strings leading from them the chemical tendencies which the genes produce. The modeling of the epigenetic landscape, *which slopes down from one's head* towards the distance, is controlled by the pull of these numerous guy-ropes, which are ultimately anchored to the genes."

Notice the focal point Waddington adopts in describing this third version of the epigenetic landscape: it is "one's head," from which the view "slopes down" into the distance. If we look closely at his description of the image, we find a curiously multiple perspective. We watch, with Waddington, as the landscape extends visually downward from "one's head" but we also feel it, haptically, as a force moving upward through those anchored guy ropes as "the chemical tendencies which the genes produce" tug the surface this way and that" (Waddington 1957, loc. 712).

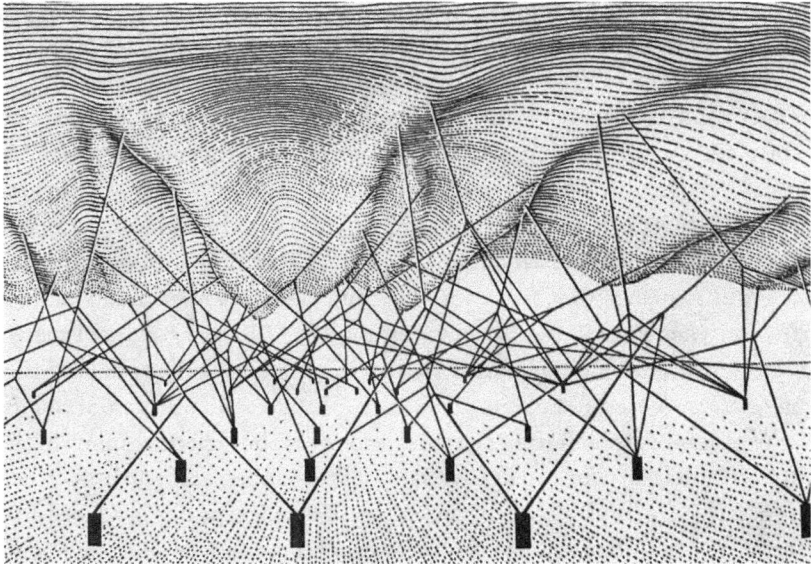

FIGURE 7.1 The third version of the epigenetic landscape captioned by Waddington, "The complex system of interactions underlying the epigenetic landscape" (Waddington 1957).

I am arguing that Waddington's third image of the epigenetic landscape incorporated a systemic view of development as a process that incorporates diverse factors and multiple directions.[4] The shift from single to multiple scales is one of the foundational principles of developmental systems theory (DST), according to Susan Oyama.[5] She argues that by moving between scales "both of magnitude and time," we can take an alternative, non-dichotomizing approach to development. As we shift "from interactions of molecules inside cells to those between persons, from the brief periods involved in the action of a hormone in the nervous system to changing relations among conspecifics over the life span, from the short-term dynamics of a population of organisms in a habitat to the slow procession of generations through evolutionary time," we gain a better sense of the relations between scales (Oyama 2000, loc. 135).[6]

Yet the issue of agency in this process of scale shifting can be challenging to grasp. To quote a friend with whom I corresponded on this issue, "Who, why, and how decides how many variables/elements to include into this new system of 'multiple scales'?"[7] The question seems to be one of ethics, at least in part, although aesthetics also figures in ways that I discuss later. One answer to that question would be to say that the choice should reflect (and is responsive to) the requirements of the context within which the choice is being made. Another answer, drawing on the work of the biologist David Noble, would be to say that there is no privileged scale and thus no necessary number of required variables or elements. Yet while reflexivity about that choice of inclusion is crucial, it is more important still to realize that these variables and elements are actually *always already included*. As Karen Barad (2012) puts it, "The very nature of materiality itself is an entanglement. Hence, what is on the other side of the agential cut is never separate from us. . . . Ethics is therefore not about right responses to a radically exteriorized other, but about responsibility and accountability for the lively relationalities of becoming, of which we are a part." The third version of the epigenetic landscape offers a more dispersed and multidirectional vision of epigenetic action, or what we might call an intra-active aspect.

Recently, drawing on the DST approach to social behavior, Iddo Tavory and his colleagues created their own version of the epigenetic landscape, which they call a "social-developmental landscape." They explain that "the trajectories in the social landscape are the developmental paths of individual subjects as they go about their lives" (Tavory et al., 314).[8] While acknowledging that their image is inspired, visually and conceptually, by Waddington's, they make four distinctions between his original epigenetic landscape and their social developmental landscape: 1) theirs is "far more open-ended"; 2) it describes "a group-level process with multiple trajectories individuals typically follow/construct"; 3) the pegs and guy ropes in their schema "are both causal and constitutive: they are not separate from the landscape itself, they make up the 'hills' and 'valleys' and are not external causal factors"; and 4) the resources/causal factors that interact are not all made up of the same "fabric" but "belong to different domains" (314–15).

The social developmental landscape has been reframed to link "genetic, epigenetic, ecological, institutional and symbolic resources," factors that shape sociocultural development, to illuminate the processes by which cultures and societies maintain themselves while adjusting to change (Tavory et al., 315). There are two crucial points about this reformulation of Waddington's third epigenetic landscape. First, Tavory and his colleagues incorporate what Waddington resisted: the potential for what one could call multidirectional agency in the epigenetic landscape:

> Like the ball in Waddington's model, individuals roll down the landscape. People, however, are not passively driven down the slopes (as the ball seems to be in Waddington's original picture): people affect their own trajectory to some extent, as they construct their social and epistemic niche—their passage is an active process which always changes it somewhat. Although individuals go through partially preexisting trajectories which they typically stabilize and "deepen," their activities may also alter the local features of the landscape with which they interact. To use sociologist Anthony Giddens's terms, there is always a "double hermeneutic" in the case of people: people are both structured by, and actively structure, social life. (Tavory et al., 315–16)

The second point, which I learned during my time at ALB, is that this double hermeneutic applies not only to the people in the social developmental landscape, but to all of the other elements, as well. To recall Barad's powerful phrase, "Matter feels, converses, suffers, desires, yearns and remembers." As I experienced the exhibits at ALB that hot day in 2014, moving between the scales of molecules and people, hormones and habitats, moments brief as a flash and as slow as the stretch of evolutionary time, I found myself appreciating how the epigenetic landscape could illuminate artistic exploration by catalyzing its own playful extension.

Founded in 2006 by the art historian and curator Regine Rapp and Christian de Lutz, a photographer, new media artist, videographer, and curator, Art Laboratory Berlin is an interdisciplinary arts space dedicated to presenting international contemporary art "at the meeting point of art, science and technology." It was one of the first spaces I visited in Berlin during the year I lived there while writing this book (2014–15).[9] I was a part of two other intellectually charged communities during the year—the Zentrum für Literatur und Kulturforschung in the autumn and the Max Planck Institute for the History of Science in the spring—yet the thinking community provided by ALB stood out to me as a remarkable third space: a hospitable matrix for intellectual development shaped for, by, and with outsiders, whether of discipline or of country. Drawing together people working in the arts and sciences, it focused on the co-created, always emergent process of thinking with and between science, technology, and the arts. Art Laboratory Berlin impressed me as an epigenetic landscape made material: a conceptual laboratory deeply shaped by the time and space within which it developed.[10]

When I first visited ALB, I knew that Berlin had been riven by the wall that divided the communist East from the capitalist West from August 1961 to November 1989. Indeed, I had dined out on a memory of our family's visit to East Berlin in the summer of 1989, only months before the opening of the wall. In addition to my husband and our two children, a German friend accompanied us, and we tried to go out for lunch on Alexanderplatz as a party of five. Yet when we entered the restaurant just after noon on a weekday and found it completely empty of customers, we were told, to our surprise, that there were no tables available for us. Staff explained that the restaurant had tables only for four or six. Our willingness to move a chair and adapt a table to our party made no difference to the restaurant manager. He told us, with a straight face, that no tables were available and refused to seat us.

To this memory of a puzzlingly inhospitable restaurant in East Berlin my unfolding experience at ALB provided a marked contrast. Housed in Wedding, one of the city's poorest and most ethnically diverse quarters that has been deeply shaped by Berlin's economic and social history since World War II, ALB is an aesthetic and civic accomplishment that should be understood in a historical and social context. After Berlin was divided in 1945 into four sectors, or occupied zones, controlled by Great Britain, the United States, France, and the Soviet Union, a further division of the city into twenty administrative areas put Wedding in the French sector and, later part of West Berlin, in contrast to its close neighbor Mitte, which was placed under the control of the Soviet Union and later joined East Berlin. Access to Wedding from the West before the wall was built was difficult, because the land around Berlin was under the control of the Russians, who had to grant permission to access it in the French sector. With the erection of the wall, access to Wedding from the West became even more controlled, with visitors required to drive a cordon of highly guarded highway through East Germany. Access to West Berlin from the East became impossible because of the wall. For obvious reasons, the commercial and industrial sectors of West Berlin dwindled away. Since Berlin was reorganized in 2001, Wedding has become part of Mitte, formerly an East German neighborhood. Just as the viability of West Berlin once relied on artists who were subsidized by the Bundesrepublik Deutschland (Federal Republic of Germany), the City of Berlin continues to support artists not only to maintain a viable population in an era of population decline, but also because the German art scene has become a tourist attraction.

This cauldron of political, economic, and social pressures has yielded the yeasty setting for artists that is contemporary Berlin. For more than a decade, a nonprofit initiative called Kolonie Wedding has brought artistic gentrification to a largely immigrant, working-class community (or, to use the language of many of the press releases, it has turned a previously down-at-heels neighborhood into a lively arts community). As the result of collaboration between one of Berlin's largest real estate companies (hoping to avoid a plague of empty apartments and storefronts) and the city of Berlin, Kolonie Wedding now supports thirty arts spaces in this sector of the city. Kolonie Wedding describes itself as part of the Creative City Berlin project (Kulturprojekte Berlin GmbH), and acknowledges support from Berlin's Department of Cultural Affairs and the Senate Department for Economy (Projekt Zukunft), which hosts an Internet portal for "the cultural sector and the creative industries."[11] Offering a well-publicized weekend gallery crawl and a series of open performances, exhibitions, gatherings, and parties, Kolonie Wedding relies on the history of support for the arts that is woven into Berlin's history. As befits the tradition of political action so integral to Berlin, the civic organization even has its own manifesto.[12]

As one of the thirty sites supported by Kolonie Wedding, ALB has been incubating transdisciplinary thinking about the arts since it was founded in 2006 as a site for collaborative work, a library for theoretical exploration, and a gallery space. Art Laboratory Berlin prides itself on its interdisciplinary origins, its orientation to specific kinds of content and a specific (and different) intended audience, the pathbreaking curatorial practices and aspirations it exhibits, its unique approach to art and the artist, and its cutting-edge understanding of the relations among art, science, and technology.[13] I discuss each of these aspects of ALB in turn.

Founded in 2006 by three art historians—Marguerite Tillberg, Sandra Frimmel, and Regine Rapp—and the visual artist Christian de Lutz, ALB became the collaborative responsibility of Rapp and de Lutz in 2009, who operate now as co-founders, co-producers, co-directors, and even co-curators.[14] To Rapp, one simple thing distinguishes ALB from the many nonprofit art institutions in Berlin: "Art Laboratory Berlin is interesting because we are *content* oriented. We want to do content production. . . . With all the different series we have, all the diverse themes and the diverse disciplines, . . . we're talking about the public that we want to nurture and feed with content." There is nothing particularly new about art galleries providing scientific content, of course. As long ago as 2002, Hal Cohen surveyed a range of galleries in the United States, England, and elsewhere in Europe where "ambitious artists and accessible technologies have modernized the marriage of biology and art into bioart [*sic*] coupling imagination and science to create animate, often interactive, works that put pretty paintings of flowers to shame." Taking its place among a small but vibrant community of bioArt laboratories—Maria de Menezes's Cultivamos Cultura in Portugal, Oron Catts and Ionat Zurr's SymbioticA Research Laboratory in western Australia, and Suzanne Anker's Bio Art Lab in New York City come to mind—ALB stresses its feminist-inflected, post-disciplinary curatorial ambitions and practices, as well as its specific way of working between modes of engaged artistic and scientific practices, to shift the relations between artist and art object and curator and those among artists, scientists, technologists, and the multidisciplinary public.[15]

The exhibitions Rapp and de Lutz have curated at ALB stretch from the "vast to the minute," from the scales of the very large to the very small—or, as they put it in the curatorial materials for their exhibit *[Macro]biologies* and *[Micro]biologies*, they have "moved from biosphere and landscape—systems, structures, creation and devastation (exhibition 1) to the level of non-human, multi-cellular beings (exhibition 2) and finally explored microorganisms (exhibition 3), as well as the minute particles or objects that form life or a basis for living (exhibition 4)." Their reference to "biologies" expresses the intention to engage the wide range of the life sciences. They solicit work "marked by a plurality of disciplines (e.g., ecology, botany, comparative zoology, biotechnology, biochemistry, microbiology, etc.)." Whatever the specialist field is that the artists concentrate on, they explain, "We are most interested in their dedication to understanding and partaking in scientific professionalism [and] in developing a hybrid field that results from the collaboration between artists and the sciences" (Rapp and de Lutz 2015, 9).[16]

I began exploring this hybrid space, on that hot July day, by strolling through those exhibitions. *[Macro]biologies, Bio-BASE: riskyZOOgraphies,* and *[Micro]biologies* were the exhibits, respectively, of Suzanne Anker, Maja Smrekar, and Brandon Ballengée, a biologist who engages in a form of citizen-science bioArt.[17] At different scales, in different media, and through different forms of practice these exhibitions exemplified the process of thinking together across different disciplines and geographic contexts about the nature of biological development, stability during change, and variation, all essential properties of the epigenetic landscape.

In a large, high-ceilinged set of rooms reached through a grassy inner courtyard across the street from the main Art Laboratory Berlin, I found *Vanitas (in a Petri Dish)* and *Remote Sensing,* two exhibitions by Suzanne Anker. Four broad tables were thoroughly covered with composed and constructed small art objects in Petri dishes (figure 7.2).[18] Some of the Petri dishes held colorful flowers, insects, seeds, skeletons of small amphibians, and even small bones from the meals Anker had eaten when she came to Berlin to install the exhibition. Others, part of *Remote Sensing,* contained delicate, exquisitely colored, three-dimensional, rapid-prototype-printed "micro-landscapes."

Anker's work is scaled in time and space, as if incorporating Waddington's understanding of the biological picture. From a contemplation of the aesthetics of life and death to the serendipity of found art and science, *Vanitas (in a Petri Dish)* explores how development stretches from the individual to the population and produces variation within continuity.[19] She compares those images, which group together "an amalgam of animal, vegetable, and mineral specimens," to "synthetic biology," a mode of scientific research that she recasts as distinctly aesthetic: "It is opening the pathways under which organisms and plants can be fabricated to create a palette of living entities not present in the natural world" (Anker 2014, 29).

In the dramatic distinction between those delicate Petri dishes full of organic found objects and the micro-landscapes of *Remote Sensing,* created

FIGURE 7.2 Three-dimensional printed "micro-landscapes" based on Petri dishes of organic found objects, part of the *Remote Sensing* exhibition at Art Laboratory Berlin (Anker 2014, 29).

by 3D printing of their digitally altered images, we can find an echo of the progression between John Piper's analogue river and Waddington's later abstracted stochastic image. Indeed, Anker describes the process she used to produce the micro-landscapes in terms that recall Ian McHarg's scale stretch from the GIS inspired plans for the Pardisan project to his close focus on the materials visible in pond water. The title of the exhibition explicitly connects it to "state-of-the-art delivered Satellite data. . . . Images [that] garner information electronically in which on-site visitations are eclipsed" (Anker 2014, 29). Yet she describes the algorithmic shaping of this micro-landscape as a metaphoric movement on the molecular scale from genotype to phenotype:

A series of rapid prototyped objects are created from Photoshop files. Images related to micro-landscapes are the result of these operatic extrusions. The software program which operates the output system makes "decisions" about color in which height, length and width are determined by programmatic aspects. Although a set program is in order, output varies and, in reality, no two specimens are identical. This then brings us to the question of code as a genotypical variable as in DNA and its variegation in phenotypical appearance. Conceptually speaking, this may be the most important issue here, in that in nature there are no identical specimens regardless of DNA equivalence. (Anker 2014, 29)

While Anker's *Remote Sensing* draws on computer algorithms to explore the emergence of variation, another exhibition I visited the same day worked with a small number of individual, material, very unprogrammable crayfish to explore the interactions between invasive and native species that led to morphological changes. That exhibition, the Slovenian artist Maja Smrekar's *Bio-BASE: riskyZOOgraphies*, represented a continuation of a longer-term project carried out in Ljubljana, Slovenia, in collaboration with biologists there. In that project, in 2002, Smrekar had built an aquarium in the style of a tent, where she housed Australian red-claw crayfish (*Cherax quadricarinatus*), a species that had just settled in great numbers in a local oxbow-shaped thermal lake, with the local crayfish from Slovenia (*Astacus astacus*). A little ladder connected the two aquarium wings, "allowing the crustaceans the possibility of crossing over and confronting each other" (Smrekar 2015, 31). At stake in that potential encounter between native and invasive species was the "juxtaposition of ecology as a scientific discipline and Ecology as one of the ideologies of contemporary zeitgeist," Smrekar has explained. "Due to the changed mechanisms of natural selection, which are caused by globalisation and which establish new relations within ecosystems, behavioural and genetic transformations in foreign as well as domestic species have become quite common. These transformations result in the changed potential of their ecological niches; as a consequence, from a long-term perspective, these species represent the greatest threat to the humankind [*sic*]of all species."[20]

For ALB, the stakes had been changed somewhat, as had the cast of characters. After consulting with Gerhard Scholtz, a comparative zoologist from Humboldt University of Berlin who discovered the existence of a

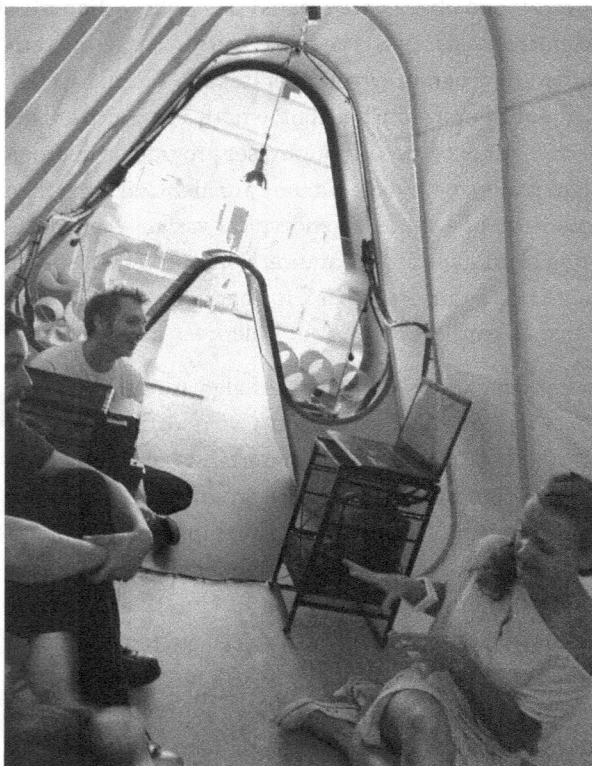

FIGURE 7.3 The A-frame aquarium tent where Smrekar installed the marbled crayfish and its near-relative, the Louisiana crayfish: a meditation on species vulnerability.

parthenogenetic variant of the marbled crayfish, Smrekar had abandoned her plan to reprise her earlier exploration of the interactions between invasive and native species. All of the invasive species of crayfish available in Berlin, she learned, were of North American origin. She decided that it would be unethical to use them in an experiment with European crayfish because they were frequent carriers of a disease that would be lethal to their European counterparts (Rapp and de Lutz, 33). Instead, at ALB she installed in her Aquatic Art Laboratory (figure 7.3) one of the workhorses of the U.S. aquarium trade, the familiar marbled crayfish, and its near-relative, the Louisiana crayfish (*Procambarus clarkia*).

That decision raised the problem of invasive species in the intimately entangled terms of epigenetics and ecology. When raised in pet shop or laboratory aquariums (rather than living in the wild), the marbled crayfish, or *marmorkrebs*, reproduces parthenogenetically (i.e., without fertilization), producing its own genetically identical clones. Because clones hold steady the genetic material of an organism, enabling one to study the other variables that shape its development, the marble crayfish has become a popular model organism in the laboratory study of epigenetics (Rapp and de Lutz, 33). It has thus joined the other, most popular animal models: the fruit fly (*Drosophila melanogaster*), the nematode (*Caenorhabditis elegans*), the chicken (*Gallus gallus domesticus*), the mouse (*Mus musculus*) and the African claw frog (*Xenopus laevis*), among others.

While of debatable value to the marbled crayfish themselves, the potential they hold for parthenogenetic reproduction is an event of no small significance for human beings. According to the zoologist Günter Vogt of the University of Heidelberg, "The greatest potential of the marbled crayfish lies in epigenetics and environmental epigenomics and in stem cell research and regeneration. The marbled crayfish also appears to be suitable for the investigation of the role of stochastic developmental variation and epigenetic inheritance in evolution and to contribute to evo-devo and eco-devo" (Vogt 2008). What this means is that the marbled crayfish could help scientists understand such issues as individual vulnerability to illness; how environmental toxins or changes in temperature result in developmental differences; the possibility of reversing nerve damage; and the probability that members of a species will develop differently given specific inherited or environmental factors. Of course the promissory nature of these scientific applications reflects the way that "sciences of the actual are displaced by *speculative forecast*" (Adams et al. 2009, 246).

Smrekar had decided to put female parthenogenetic crayfish (Virginales) in the A-frame aquarium together with the male Louisiana crayfish. Would the Louisiana crayfish climb up the ladder to breed with the female (but parthenogenetic) marbled crayfish? The significance of this work of art was tied to biological process as much as any scientific experiment or project in citizen science. On the steamy day in Berlin when we curled ourselves into the space behind the A-frame plastic aquarium tent to listen to Smrekar talk about the work, we learned that things had unfolded in ways that were both unpredictable and as yet inexplicable.

The population of Louisiana crayfish had dropped dramatically. However, in the other tank the marbled crayfish had lost only one adult and had gained a fairly large number of juvenile crayfish. Only at their maturity could it be determined whether they were parthenogenetic. The observation with which Smrekar concluded her presentation linked this organism-level art project to its broader population-based implications: "Invasive species grow exponentially and then crash, which is what makes humans an invasive species." From this A-frame aquarium, which reminded me as I sat there of the conical structure of the epigenetic landscape itself, the likelihood of continued species development—*all species*—seemed tenuous.

Unlike Günter Vogt, who views the marbled crayfish as a being subject to human operationalization and exploitation, Smrekar's perspective is intra-active: the environment shaping the crayfish, she suggests, is the same one that was shaping us as we watched and considered its transformation. All of us, as we sat within that A-frame aquarium, were invasive species, learning the lesson of what Barad calls "different kinds of causalities": "I really want us to specify more carefully the different kinds of causalities, and how to think causality again. And that is partly what I mean by the notion of "intra-action" as proposing a new way of thinking causality. It is not just a kind of neologism, which gets us to shift from interaction, where we start with separate entities and they interact, to intra-action, where there are interactions through which subject and object emerge, but actually as a new understanding of causality itself" (Barad 2012).[21]

The final exhibition I visited that day was by Brandon Ballengée, one of the first practitioners of a form of bioArt as citizen science. It thus took a more direct and embodied approach to exploring the vagaries of biological development. Incorporating the affective potential of art and science, lay engagement and expert knowledge, the exhibition raised my spirits simply by its recourse to environmental activism. A herpetologist and artist, Ballengée has extended epigenetic concerns into the realm of ecology and environmental activism. His exhibition at ALB presented a series of video installations drawn from his ongoing work as an artist and scientist as he tracked, and created art about, the worldwide rise in amphibians with developmental abnormalities. One video, *Reliquaries*, showed photographs of frogs that had died from their bodily deformations. The photographs were exquisite: linear and graphic, the creatures had been cleaned, stained, and then scanned at a scale chosen by the artist "so the frogs appear approximately the size of a human toddler, in an attempt to invoke empathy in the viewer instead of detachment or fear: if they are too small they will dismissed but if they are too large they will become monsters."[22]

Another video documented the eco-activist project Malamp, or Malformed Amphibian Project. Since 2002, Ballengée has been taking students and members of the public on field trips to serve as citizen-science volunteers, conducting research and amassing data that enable them to gauge the health of an ecological system. This video, screened on monitors lying on the gallery floor as if they were windows into an underground river, revealed the extent of the environmental damage caused by an oil spill. In its attention to the microscopic beings in the water, whose developing deformities register the collapse of an entire ecology, the project incorporates the central figures from the first and second versions of the epigenetic landscape: the river and the developing embryo. In the social practices essential to its production, the project expresses the attention to developmental systems, scaling up from the microscoping developing beings to the macroscopic development of the citizen-science volunteer community. Ballengée, who does not separate the two intellectual and epistemological communities in which he works but, instead, uses them to provoke continuing growth in each other, is enacting the convergence of in vivo and in vitro. As Ballangée explained to Paulette Beete, who writes a blog for the National Endowment for the Arts:

> When I am doing the science, when I'm doing the lab work or the fieldwork, I start thinking about art projects. When I'm having these experiences, like literally holding a particular frog with a deformity or fish or animals in these ecosystems, my brain starts thinking about art and visuals. I want to create things to describe this experience, to give visual form to this experience. And when I'm making the art, my brain starts thinking in other directions, like what if I do this experiment because it may show us this? My art practice and science work really inform and inspire each other, so it's literally cross-pollination, where I couldn't do one without the other.[23]

While Anker's bioArt plays with the algorithmic nature of development as modeled by a landscape in a Petri dish and Smrekar's explores the "risky ZOOgraphies" of the clash between native and invasive species, sexual and asexual reproduction, Ballengée's biologically grounded citizen science follows the river to document development gone awry. As fieldwork carried out in Canada, the United Kingdom, and the United States, and as a bioArt exhibition that I encountered in Berlin, Ballengée's environmental art and science united all three versions of the epigenetic landscape as joint interventions linking the life sciences and the arts.

Less than a month later, during a two-day workshop on synthetic biology at Art Laboratory Berlin led by the British artist Howard Boland of the artist collective C-Lab, in cooperation with Desiree Förster and Daniela Silvestrin, I had an encounter with what I came to think of as pre-fab biological development. The field of synthetic biology is a complex one, its potential both too far-reaching and fraught and the speed with which it is developing too breathtaking to do more than stay at the level of my own (novice) experience. When I went online to the Registry of Standard Biological Parts (the iGEM Registry), I found it offering the "Freiburg 2013 CRISPR/uniCAS," one in the series of methods for germline DNA modification that in 2015 was raising bioethical and social alarm. Now, as this book is going to print, it has received a regulatory green light in the United Kingdom.[24] The Synthetic Biology Workshop was my introduction to the Bio-Brick/iGEM world, whose website includes not only a catalogue and repository of standard biological parts but also protocols, instructions, and the material implements of laboratory work.

As participants learned when we arrived, we would for the next several days be working with what are called "Bio-Bricks": pre-packaged biological structures needed for genetic engineering that are made available through the Bio-Brick Foundation's iGEM Registry to registered teams working in the Synthetic Biology Network.[25] Among the fifteen or more people who joined me in that shared liminal, virtual, conceptual, and practical space were tissue-culture artists, musicians who played electrode-implanted plants, a founder of the European television channel ARTE, an artist from New York City, a psychologist from Pennsylvania, and a stem cell biologist who organizes the Science Hack Day in Berlin. In this vibrantly divergent community, we would—the informative handout explained—"be performing an experimental mixing of genes . . . producing genetic biosensing devices that drive expressions of fluorescent proteins from three different colour spectrum [sic]."[26]

Yet as we were guided through the processes of polymerase chain reaction, gel electrophoresis, ligation of DNA, and transformation of DNA, what most of us really learned was the importance of time, failure, and intuition in laboratory work. We learned how to multitask while the PCR processed in the thermal cycling machine and later as our gel trays set in their own time; just what we *could* learn from experiments that did not work as expected; how to trust the intuitions that pull you toward the techniques that are most helpful; and that our only hope for getting the sequence to flow smoothly was to have a system for every step of the procedure. To give just one example of the granular lessons I learned that day, consider the difficulty of pipetting into a gel tray without puncturing the wells (figure 7.4). Eight months later, reading my notes on the experience, I am plunged once again into the affective education that laboratory work provided: "So—the experience of filling the well was terrifying. Instructions (from Howard): breathe out, be happy, steady yourself, and when you're feeling happy, slowly put in the pipette. Then, when it's flowing in, let it flow until it's all released, and slowly draw up the pipette. Well—easier to say than to do. As the pipette was coming up, I brushed the walls of the agar, and I may also have punctured the walls. Abject failure: so not meditative!"

The second day of the workshop took place in the laboratory of the Technical University of Berlin. There, led by some of the undergraduate and graduate scientists and engineers who participated in the International Genetically Engineered Machine (iGEM) competition, we pipetted out the cell media that had come from the centrifuge, attempting to do so without losing the pellet of plasmid DNA at the bottom, which we then injected into a Petri dish to incubate for sixteen hours. By the end of that day, the flow of technical information had peaked, and my ability to incorporate it was waning. My notes for the day are a mixture of numbered lists indicating the order of the procedure and two boxed words announcing in large capital letters the final step in the day's procedures: PATIENCE! and GO TO THE BEER GARDEN!

The final morning of the workshop, we met once again in Wedding at ALB, where we gathered in front of a video screen linked to a projector streaming straight from the laboratory to find out whether we had been able to grow our fluorescent bacteria. The results were mixed: the Petri dishes that had been inoculated by the stem cell scientist and the T-cell

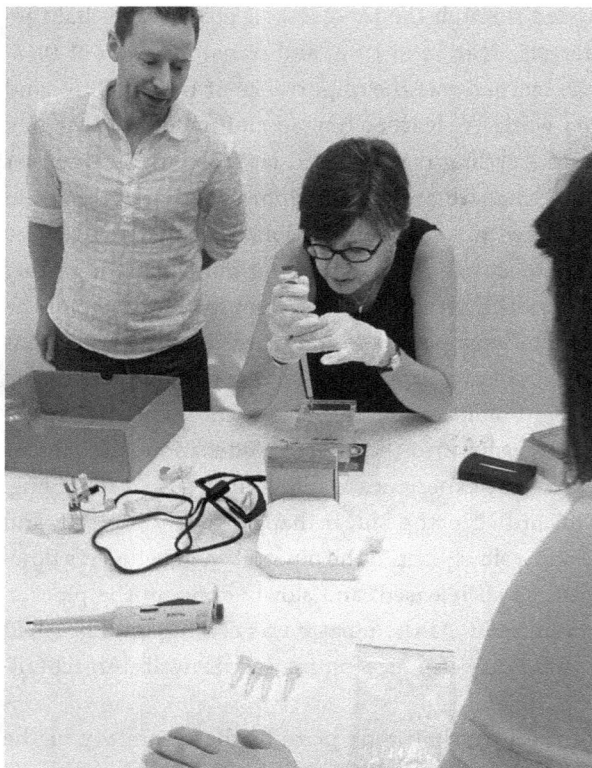

FIGURE 7.4 Learning to pipette with the iGEM team at Art Laboratory Berlin. Photograph courtesy of Gowen Roper.

researcher glowed vibrantly green, the dots scattered across their surface with the few red "tester" inoculations also visible. Some other collaborative teams had succeeded in raising some glowing colonies. But the rest of the plates showed only the tester spots. We had been failures at synthetic biology.

Yet as I thought about the three days spent with this group of scientists and artists, that failure seemed as catalytic of growth as any other kind of scientific failure. At first I wondered whether we had been engaging in performance art. Yet as I thought more about that, it seemed to me that the spectacle and "to be looked at-ness" of performance art was not exactly what we had all experienced. Rather, it seemed that we had participated in a third kind of learning, a *tiers instruit*. Traveling across our various disciplinary perspectives, taking the opportunity to be wrong, we created a new collaborative way of thinking and working with biological development—indeed, a new response to synthetic biology.[27] As my notes on the project reveal, I felt a desire emerge to talk back to the core concept of synthetic biology: the incorporation of a machinic model for the creation of life or, to put it another way, the reliance on pre-formulated genetic materials to assemble something biologically new. I find, scrawled in my blue journal, the following: "What about the unexpected? Does Synthetic Biology give itself space to be emergent?" This question stayed with me as I considered the difference between synthetic biology and the kind of synthetic post-disciplinary engagement I found during the rest of the year at ALB.

While the Synthetic Biology Workshop tested the limits of my skill in the laboratory, and troubled me by its formulation of both variation and failure as experiences to be avoided rather than learned from, it did not diminish my interest in the possibilities raised by bringing artists, scientists, and scholars together to think about development. Later in the year I visited ALB again to attend the show *The Bacterial Sublime*, by the British bioArtist Anna Dumitriu. Textiles, as demonstrations of the delicate webbing of the world, were the core of the show, which presented Dumitriu's works over the previous decade. Currently the self-proclaimed head of the ironically named "Institute of Unnecessary Research," Dumitriu has been investigating the role played by bacteria in connecting the realms of medicine, human health, textile artistry, and communication.[28] Her beautiful show at ALB linked the prokaryotic and eukaryotic worlds, as well as the different social and epistemological worlds of art, biology, and medicine, as if Dumitriu, too, were enacting the third version of the epigenetic landscape, this time by focusing explicitly on the delicate webs, those guy wires linking the genes (and their chemical signals) to the landscape itself. In one corner, a crocheted bedspread lay draped on a chair, the work still incomplete, a crochet hook beckoning visitors to continue the project with crocheting of their own.

The volume documenting this exhibition, produced by Rapp and de Lutz, includes photographs of the wide variety of material objects whose surfaces, internal weaves, and conceptual frames Dumitriu enrolls to explore the role bacteria have played in human health. I gaze at the photograph of a tiny wire hospital bed and screen from the miniature tuberculosis clinic:

its blue-and-white striped ticking mattresses are stained with the dyes that in the past were used as remedies for tuberculosis (TB), and its little pillow is "impregnated with the killed extracted DNA of TB" (Rapp and de Lutz, 14). I marvel at the "altered antique era romantic maternity dress" of a beautiful rose, its pale silk impregnated with killed Mycobacterium TB and dyed with extracts of walnuts and madder roots, once thought to be remedies for the disease (18). I notice the explicit attention to the gendered impact of Dumitriu's version of the epigenetic landscape. The beautiful fabric of the maternity gown hides a more malevolent tissue of causes and effects, as Dumitriu reminds us: because of TB's reputation as a heritable disease, men and women with TB were once forbidden to marry, and pregnant women with TB were required to have medical abortions (18).

I pause the longest at the close-up photograph of tiny gray felted lungs, traced with tiny white stitched veins that Dumitriu has crafted to reveal the stages of the disease and different kinds of treatment for TB. Made of a compound of wool and the material we would in the United States call "dust bunnies," floor sweepings with their dangerous burden of dried human spit coughed out by the tubercular inhabitants, these tiny, butterfly-like gray-brown objects piled together in the installation "Where There's Dust There's Danger," miniaturize, domesticate, and aggregate the individual impacts of this disease. Dumitriu's intensive use of women's handiwork such as weaving, sewing, dyeing, and embroidering to explore the meanings of health and disease in a social context offers a marked alternative to the notion of women as vectors that is found in much contemporary epigenetic research (Richardson 2015). Rather than introducing toxins that disrupt growth, women in this art project enhance intellectual and emotional growth. Like the crocheted coral reefs of Margaret and Christine Wertheim's Hyperbolic Crochet Coral Reef Project at the Institute for Figuring, Anna Dumitriu's textile art engages in a multiscale developmental process we might term Sym-poiesis, or "multi-species-becoming with, multi-species co-making, making-together, *sym-poiesis* rather than auto-poiesis."[29]

In this chapter, I have argued that, in its curatorial and exhibition practices, ALB exemplifies the systemic, interactive view of developmental processes expressed in Waddington's third epigenetic landscape and therefore that the epigenetic landscape can illuminate artistic practices. This should perhaps not surprise us, since the reverse certainly is true. As Donna Haraway recalls about studying with the ecologist and zoologist Evelyn Hutchinson, the author of "Circular Causal Systems in Ecology":

> He had us reading C. H. Waddington and pondering the importance of painting in his thinking about genetic assimilation and developmental plateaus. Our thinking about ecology, evolution, and genetics in the 1960s, when I was a graduate student, already involved some serious consideration of modernist painting, as well as modernist poetry. We talked about all this in our biology tea groups; our lab meetings involved thinking about philosophy, biology, art. I was really lucky to be part of an educational scene that took for granted that artistic practice is intrinsic and necessary to good thinking and to good science. (Haraway with Kenney 2016, 263–64)

The crucial word, of course, is *intrinsic*.

Anastomosis

This study has traversed the epigenetic landscape looking for new vantage points on the complex and contested scientific field of epigenetics. The perspectives offered by embryology, graphic medicine, landscape architecture theory, and bioArt have opened up some of the ways that figure expands our understanding of development at multiple scales.[1] Inherent in all of these perspectives has been a feminist commitment to understanding how "what counts as nature" is something we shape together, every day, in our ideas and our actions, as in our choice of the figures with which we communicate about it, figures such as the epigenetic landscape. Now I want to leave the figure behind for a moment to talk about the feminist implications of the field of epigenetics itself.

In the introduction, I discussed how researchers have faltered in their assurance that epigenetics can explain the causes of diseases through epigenome wide association studies (EWAS). Feminist scholars, too, are challenging the promise that some once thought this field held.[2] The historian of science Sarah S. Richardson makes a convincing case that epigenetics is being narrowed and redirected away from its potential for feminist resistance. Surveying the uses of the field of epigenetics in research on sex and gender differences, she finds that even potentially paradigm-shifting findings are often reframed to endorse existing conceptions of sex and gender as binary, programmable, and stably retained over time. She reports that even when the results of an epigenetic study of sexual differentiation in rat brains (Nugent et al. 2015) seemed to affirm gender plasticity and phenotypic diversity, it was instead reframed by the authors of the study and the scientific community as confirming "the paradigm of hardwired and dimorphic sex differences in the brain" (Richardson, forthcoming). This reframing of the implications of epigenetics coincides with the persistence of a long-standing explanatory framework that privileges a linear, gene-centered, and "programmed" approach to development.

What this means, she argues, is that epigenetic research is being directed to 1) providing "a new toolkit for elucidating the molecular mechanisms of hormone-gene regulation"; and 2) elucidating the black-box causes of effects already taken to be overdetermined within an accepted model of sexual dimorphism, such as the timing of puberty and the development of homosexuality (Richardson forthcoming). The effects Richardson identifies in this study are familiar, if discouraging. As Aaron Panovsky (2015) has demonstrated, even affirmations of the vaunted complexity, nonlinearity, and potentially paradigm-shattering implications of epigenetics research represent not so much an investment in this new explanatory model as a set of rhetorical holding actions and coping strategies to reorient the program of epigenetics research and compensate for the failings of the molecular genetics model and its underfunded research programs.

I described in the introduction how epigenetics has been welcomed for its promise to illuminate multigenerational developmental impacts, offering a new tool with which feminist and antiracist researchers could document the slow violence whose temporally and spatially dispersed effects otherwise escape measurement. The hope is that research on epigenetics promises to do more than explain the onset of illness in an individual patient: it could be used to model the multiple variables shaping the health of a population, or an ecosystem. But such eager promissory proclamations participate in the hype that makes epigenetics problematic for feminists; the stress here falls on "could." Despite the seeming promise of epigenetics to affirm plasticity, multiplicity, complexity, and variety in ways likely to delight feminists, the development and implementation of this field of knowledge is itself subject to its own form of environmental constraints and costs, reflecting the range of disciplinary fields within which it is deployed. Yet as Richardson reminds us, "In each scientific field, the explanatory reach of epigenetics emerges temporally through local practice and contestation and against the backdrop of received intellectual frameworks. Epigenetics is not only a material mechanism, but also a fluid imaginary functioning diversely across heterogeneous social spheres."[3]

In a story by Jorge Luis Borges, a man discovers a corner of his cellar through which he can see everything in the world. The man tells a friend, who ventures down into the cellar to corroborate the claim. After a while, the friend climbs back up the stairs. He saw nothing, he says; he should take it easy, he tells the man. He needs a rest. But to the reader, the friend confides that he has seen the Aleph. "What my eyes saw was *simultaneous*; what I shall write is *successive*, because language is successive. . . . The Aleph was probably two or three centimeters in diameter, but universal space was contained inside it, with no diminution in size. Each thing (the glass surface of a mirror, let us say) was infinite things, because I could clearly see it from every point in the cosmos" (Borges 2015, 201).

When I began writing this book, I thought of the epigenetic landscape as my Aleph, an image that illuminated everything in the world. I am still inspired by its remarkable conjunction of simultaneity and sequence, order and disorder, vastness and minuteness, art and science. In fact, such opposed terms falsely separate qualities that are inherently entangled (Barad; Braidotti; Haraway). Yet the process of writing about the epigenetic landscape has been one of paring down, paring down, paring down. In the end, I have focused on the image itself, in its three variations: the riverine landscape; the egg cell or embryo on an abstract hill; and the view from the underside, guy wires and all. Beginning by asking how C. H. Waddington's different intellectual and affective environments shaped the model as it changed over time, I have traced how its signal qualities made a difference as they were put to work in some very different and heterogeneous social spheres—embryology, comics, landscape architecture, and bioArt—to access a more complex understanding of biological development at multiple scales.

While writing this study, I was often asked why we should learn about epigenetics or the epigenetic landscape unless we are scientists ourselves; particularly, why should those of us who are feminists be interested in epigenetics? The answers I gave then I still believe: that we should learn about epigenetics so we are able to contest the way the field has been and is being redirected and narrowed in scientific research and medical practice; that we should do so to recapture the potential of the epigenetic landscape as a methodological prompt crafted at the intersection of art and science that can, when used creatively, amplify the options we have for exploring the complex network of interactions that is biological development; that we should do so because over its developmental history, the field has seen many promising interventions that were resisted, reinterpreted, redirected, or refused and to which we should return; and because even in this frequently reductive and deterministic postgenomic era, feminist scientists are hoping that the epigenetic landscape will enable them to forge a more multidimensional science. But there is one additional reason to learn about epigenetics and the epigenetic landscape: "Das ist ja toll!"

The phrase is, of course, from the interchange between the embryologists Christiane Nüsslein-Volhard and Eric Wieschaus, when, during their work on the development of Drosophila embryos in the late 1970s, they discovered a larva of the fruit fly with an underdeveloped ventral side. The lore has it that Nüsslein-Volhard saw the strange larva, commented, "Das war ja toll!" and named the mutant gene "Toll." Ten years later, the two of them were awarded the Nobel Prize for Physiology or Medicine for their discovery of a group of fifteen genes that control embryonic development in flies, whose mutations would cause marked developmental deformations. The press release on the Nobel paints a memorable image of the two scientists, sitting together at a special microscope that enabled them both to look carefully at the same embryo at the same time and working their way slowly, over the course of a year, through a long series of embryos until they had assembled their findings.[4] But more memorable than that is the complexly untranslatable word with which Nüsslein-Volhard announced the finding. Since then, commentators on this episode have translated "Toll" variously as "great," "weird," "amazing," "curious," "crazy," and even "droll." The charge of the term is mostly positive, but as Mike Fortun has pointed out, it has an element of the uncanny, as well, "reminding us of the capacity of any semiotic network to flip over into a near-opposite effect, just as the strange and the familiar suggest, imply, or produce each other" (Fortun 2015, 33; Hansson and Edfeld 2005, 1085).

The story of Nüsslein-Volhard's work, which brought back into focus the previously ignored cytoplasm, shifted the focus of research back to before fertilization and led to a gradual movement away from the discourse of gene action (the notion that genes determined and controlled all biological development). This story has been told elegantly, and there is no need for me to relate it again (see Fortun 2015; Keller 1996). Suffice it to say that Nüsslein-Volhard's work on the role of the maternal contribution through the cytoplasm in Drososphila embryos reoriented our understanding of development in ways consonant with the insights of epigenetics.[5]

For me, the phrase "Das is ja toll!" not only recalls Nüsslein-Volhard but also brings vividly back into mind my first months in Berlin in 2014. In the mistaken notion that I would be able to learn German with the same (relative) comfort that years earlier I had studied French and Spanish, I signed up for a month of intensive daily German classes at the Goethe Institute. The classes were immersive, and the complexity of the grammar and vocabulary was stunning to someone used to Romance languages. A vivid memory of the early days in that class was puzzling over the meaning of the word *toll*. Did it mean "great," as we were sometimes told? Or "weird?" Or even "crazy"? For me, *toll* captures the joy, disorientation, serendipity, and intensity I experienced in a country whose language was not my own, writing about epigenetics and the epigenetic landscape. So, indeed, I would add to my list of reasons to learn about epigenetics and the epigenetic landscapes: because to do so is indeed a *toll* pursuit.[6]

Three brief examples of redeployments of the epigenetic landscape in the current work of one feminist scientist may convey just what is *toll* about using this evocative figure to think about development from a feminist perspective. The biologist Anne Fausto-Sterling has spent her distinguished career challenging the narrowness of sex/gender research. The author of numerous books and articles on gender, sexuality, race, and science, she may perhaps still be most widely known for her article "The Five Sexes," which appeared in a shortened version in 1993 in the *New York Times*.[7] There, she critiqued the rigid two-sex model, charting the oppressive treatment of intersexuals by medical doctors such as Hugh H. Young, who taught both hormonal and surgical ways to make people born intersex into one sex or the other. In a later, revised version of the essay, she argued that sex and gender should be thought about not as a binary, but "as points in a multidimensional space" (Fausto-Sterling 2000, 22).

Fausto-Sterling began as an embryologist, drawn to the complex choreography of multiscaled layers of development the discipline revealed. As she told an interviewer in 2015,

My original start was as an embryologist in developmental biology. . . . I took an embryology—a development—course, and I saw a 1930s film of gastrulation. It basically blew me out of my chair. I said, "This is what I've got to study." It was so phenomenal watching these cells flow in sheets and turn themselves into these three-dimensional topographies where they basically refolded themselves from a ball into a gut tube and then a mesoderm and then an epidermis and a neural tube and all of the things that went to form the basic body axis.[8]

The scene she describes might bring us back to Nüsslein-Volhard and the serendipity of discovering how the Toll gene mutation reshaped the basic body axis. Or it could call to mind the rhythmic movements whose graphic representation was a strategy shared by fencing instructors, botanists, and embryologists (Wellmann 2017). But writing now, it brings to mind Christopher Niemann's origami rose on the cover of the *New Yorker* magazine, enacting a process of rhythmic folding that becomes animation, *that brings the rose to life.*

We have seen that Waddington and his colleagues at the Strangeways Research Laboratory, embryologists and tissue culturists, used animation to enroll public interest in their research field. So it should not surprise us that Fausto-Sterling recently has turned to animation of Waddington's epigenetic landscapes to help her think about sex and gender differences in infants and children. Working now in dynamic systems theory, a field bridging systems biology and developmental psychology, she focuses on the balance between stability and change in a constantly moving system.[9] From this perspective, she understands development as occurring not simply through the mixture of genes and environment but, rather, as a process of *iterative* change, a spiraling elaboration "that works with whatever the body has at hand" (Fausto-Sterling 2015).

To that end, working with the Brown University undergraduate Zachary Silverberg, she created an animated comic version of the epigenetic landscape to model the development of gender differences dynamically in infancy.[10] Five babies dressed in colorful infant clothes tumble down a sloping landscape from birth to eighteen months. The infants are brown, pink, tan; they are indeterminately sexed and gendered; even at the conclusion of the animation their gender and their sexuality do not map onto an easy binary but instead distribute across a field of pink, purple and blue, including one possibly laggard, gender-indeterminate child who peeps from behind one of the hills. The animation moves the analysis of development from the prenatal to the postnatal period and reveals that the landscape, too, is reshaped, as is the individual's physiology: "A key concept is that, rather than arriving preformed, the body acquires nervous, muscular and emotional responses as a result of a give and take with its physical, emotional and cultural experiences" (Fausto-Sterling 2015). Fausto-Sterling's goal is to drastically reconceptualize the notion of socialization, she jokingly told "all the feminists in the room" at a recent keynote speech to sex/gender researchers. The animation helps us see it as the application of outside pressure not on docile material but, rather, on a vigorous two-way process.[11]

In April 2016, I asked Fausto-Sterling why she incorporated the animation in her presentation:

> I give these talks and I'm trying to get people to let go of what was nature and what was nurture. I read all of the old Waddingtonesque things from Esther Thelen's work, and I still found that when I was using PowerPoint I was waving my arms around a lot with those and it was still very static. I found that when I was trying to explain the big view, I was trying to explain a story about gender and where it came from. So the cartoon animation was the first one that I did actually—a student did the animation, but I guided him; I pushed him to develop the concepts. He came up with the idea of having the one child end up stuck behind the hill. . . . At first it was all very static. . . . The kids don't stay in the same place. I said you have to make both sides of it active, and then he ended up with this one kid behind there, behind the hill, and I said, "That's actually great. That's a great idea." So we left it that way. I gave him the timing on it and the actual ages at which roughly empirically we know these things were happening.[12]

Representing the kind of scientific cross-fertilization Jan Baedke described as integral to the uses of the epigenetic landscape for transdisciplinary conversation, Fausto-Sterling, trained as an embryologist, is now drawing on the work of the developmental psychologist Esther Thelen and her colleagues. They approached development as multimodal and dynamic, adapting Waddington's epigenetic landscape to express "the multicausal, fluid, contextual, and self-organizing nature of developmental change . . . and the role of exploration and selection in the emergence of new behavior" (Thelen 1995, 79). The animation of the epigenetic landscape focuses not on the development of the embryo in a system of gene interactions, but—in the Thelen tradition—on "growing humans as true dynamic systems" (93).

This study has argued that the epigenetic landscape has served as a valuable resource for fields as diverse as cartooning and medical communication, landscape architecture, and bioArt. Fausto-Sterling's model also incorporates the modes of working with the epigenetic landscape that we have seen in those realms beyond the life sciences, particularly the strategies of art and design, serial visual display, and ecological attention. Note, too, the kinetic, affective, and intra-active processes that Fausto-Sterling and her collaborator have incorporated into their modified epigenetic landscape. The behavioral development of the infant is represented by the changing topography and colors of the landscape plane, so that changes in the environment are both triggered by and reflect changes in movement and behavior. By integrating images of complex processes into a narrative that can be immediately apprehended, Fausto-Sterling uses the animation as an analytic resource. As she explained, "I'm trying to use dynamic systems visualizations to put them all together on a single piece of paper and make a plausible story. So I see the mapping as a hypothesis or a set of hypotheses."[13]

Fausto-Sterling shared with me by e-mail the scanned image of another epigenetic landscape that she also hopes to animate. I include the lengthy comments she made on that image during our Skype conversation because they show how she is using the epigenetic landscape as space for innovative feminist scientific work:

> I have one other image that I want to animate, and it's a Waddington image; I'll send it to you. It's one where the landscape is tethered to genes. It's the view from underneath, but I've combined that with some images of spectrum of gender identity or gender expression, and what I would love to do is, instead of having those genes hold the landscape in place, I've tried to reconceptualize them as weights that swing back and forth and make the landscape undulate. And they'd be for me everything that the birth physiology of the infant is, to what the parents' behavior is, to a whole bunch of contributing weights that are shaping the landscape.
>
> *One thing that doesn't work, that isn't accurate in the use of the Waddington imagery is that you can't see the landscape being changed. You still*

have to say that the landscape is being changed. So, I'm trying to take this to the Waddington images that are so evocative and I've always loved since when I was a much younger scientist and turn them to my purpose.[14]

By adapting the epigenetic landscape, Fausto-Sterling transcends the limited, one-way understanding of development revealed in Waddington's Serbelloni paper (his response to the chick hatched from the world egg). Instead, turning the Waddington images to her purpose, she is creating an intra-active model of gender identity development.

Fausto-Sterling is currently working on animating a third image whose debt to John Piper's inaugural image of the epigenetic landscape is inescapable. She describes it as "a river bed aerial view for an academic publication on gender and motor development." Here again, she explained, the turn to animation is an essential part of her thought process: "I needed to visualize how maternal input fits into everything else we know about what influences motor development."[15] The image is a blurry one: across a set of winding, wandering, interweaving streams that look like something from the Mumbai estuary or the Mississippi delta we see what appears to be a flow chart, winding upward from Prenatal to Birth to Five Years to Adult Motor Function, overlaid with an upwardly spiraling set of factors: Neural Tube Development, Infant Variability, Class, Culture, Caregiver Behavior Patterns, Dyad Interactions. A two-way interaction of rainbow arrows reveals that, underlying Infant Variability and contributing to it via a thick green, leftward-curving, upward arrow are two interconnected sets of events: "Motor Activity Stimulates Neural Development" and "Neural Activity Bursts Stimulate Motor Activity." A question in blue hovers to the right of the river: "Maternal Physiology and Fetal Hormones Modulate Neural Responses?"

What is Fausto-Sterling working with in this image? She explained in an e-mail that she is "especially interested in the anastomoses and room for variability allowed in the aerial image."[16] Later, over Skype, she elaborated on that point: "Aerial photographs of rivers provide me at least with enough of a starting point. Anastomoses of tributaries and sandbars which allow you to build into it a variability of input and a variability of trait formation."[17] The idea of using aerial photography to get a different perspective on development was familiar. I had seen the role it played for Waddington and Piper, as well as for Mathur and da Cunha. But the term "anastomoses" was new to me, so I looked it up. *Toll!* I was delighted to find that it linked many of the disciplines across which I have wandered in this study: "An anastomosis (plural anastomoses, from Greek ἀναστόμωσις, communicating opening) is the reconnection of two streams that previously branched out, such as blood vessels or leaf veins. The term is used in medicine, biology, mycology, geology, geography and architecture."[18]

The flowing process of divergence and reconnection that is "anastomosis" could be a metaphor for my own theoretical and practical commitments in this study. Just as Fausto-Sterling works with the affordances of the river image, its anastomosis, as she adapts the epigenetic landscape to analyze the plenitude of influences shaping gender development, I have been inspired by the river figure to explore the fluid imaginary that is the epigenetic landscape. Just as "anastomosis" as a term figures in a wide range of fields, the meanings of the epigenetic landscape—as they have diverged and reconnected—have drawn me into the fields of biology, geology, geography, landscape architecture—and perhaps even mycology, if we recall the beautiful tiny felt lungs in Anna Dumitriu's exhibit at Art Laboratory Berlin, testament to the agency of fungi and bacteria in the story of tuberculosis. Along the way, I hope to have demonstrated the importance of the epigenetic landscape to model an understanding of development that is sensual, visual, and narrative, as well as mathematical (Haraway with Kenney 2015, 257).

NOTES

⌄

Introduction

1. This is the most accessible definition, being the first hit on a Google search for "epigenetics" (https://www.google.com/webhp?sourceid=chrome-instant&ion =1&espv=2&ie=UTF-8#q=epigenetics). Wikipedia offers another: "In the science of genetics, epigenetics is the study of cellular and physiological phenotypic trait variations that result from external or environmental factors that switch genes on and off and affect how cells express genes" (https://en .wikipedia.org/wiki/Epigenetics). Of course, the complex history of this term escapes these definitions, as I discuss later.

2. There has been a nearly 300 percent rise in publications on the topic between 1997 and 2008, according to GoogleNgram (https://www.google.com/webhp ?sourceid=chrome-instant&ion=1&espv=2&ie=UTF-8#q=epigenetics, ac-cessed July 26, 2016). After the definition, Google performs its own version of the hype when it gives an example of the word used in context: "Epigenetics has transformed the way we think about genomes."

3. I draw heavily here from Robertson 1977, as well as from the acceptance speech by Waddington's daughter Dusa McDuff in January 1991, when she received the Satter Prize of the American Mathematical Society in San Fran-cisco. McDuff, in turn, has also been elected a Fellow of the Royal Society.

4. For discussion of Waddington's Strangeways years, see Squier 2004.

5. See http://www.grahamstevenson.me.uk/index.php?option=com _content&view=article&id=643:barnet-woolf&catid=23:w&Itemid=124, accessed August 8, 2016.

6. McDuff has also recalled that her mother was thought to be "unusual" because she had a career as a highly talented architect and was forced to settle for a "civil service job since that was the best position which she could find in Edinburgh" (http://www.math.stonybrook.edu/~tony/visualization/dusa /dusabio.html, accessed August 10, 2016).

7. "'It seems very pretty,' [Alice] said when she had finished [Jabberwocky] 'but it's rather hard to understand!' (You see she didn't like to confess, even to herself, that she couldn't make it out at all.) 'Somehow it seems to fill my head with ideas—only I don't exactly know what they are!'" Humpty Dumpty's response is worth citing in full, because it has become a literary classic: "'Slithy' means 'lithe and slimy.' 'Lithe' is the same as 'active.' You see it's like a portmanteau—there are two meanings packed up into one word'" (Aliceinwonderland.net, http://www.alice-in-wonderland.net/resources/chapters-script/through-the-looking-glass/chapter-1, accessed May 10, 2016).

8. According to Bateson's obituary, his other favorite references were "William Blake, Samuel Butler, Lamarck, Alfred Wallace, Darwin, C. H. Waddington, R. G. Collingwood, Whitehead, Russell, the Bible, St. Augustine, Von Neumann, [and] Norbert Wiener" (Levy and Rappaport 1982, 379–94).

9. As Waddington explained in 1947, "I introduced the word 'epigenetics,' derived from the Aristotelian word 'epigenesis' . . . as a suitable name for the branch of biology which studies the causal interactions between genes and their products which bring the phenotype into being" (Waddington 1968, 9, cited in Jablonka and Lamb 2014, 393). Waddington went on to coin another term, "epigenotype," which, Scott Gilbert (2012, 21) explains, means something like "the interactive developmental genetic toolkit." This term fused epigenetics with genotype because, as Waddington (2012, 10) explained, "We certainly need to remember that between genotype and phenotype, and connecting them to each other, there lies a whole complex of developmental processes."

10. As Baedke points out, Waddington did modify the epigenetic landscape in response to the discovery of the *lac operon* and the developmental model it illuminated: "he reinterpreted the [epigenetic landscape] in terms of an attractor surface backed up by a group of topological approaches, especially catastrophe theory" (Waddington, 1969, 1974, 1977a; Baedke 2013, 759). The Wikipedia definition tellingly stresses the genetic perspective, in a potent testament to the role of the discovery in sidelining an epigenetic perspective: "lac operon (lactose operon) is an operon required for the transport and metabolism of lactose in Escherichia coli and many other enteric bacteria. Although glucose is the preferred carbon source for most bacteria, the lac operon allows for the effective digestion of lactose when glucose is not available. Gene regulation of the lac operon was the first genetic regulatory mechanism to be understood clearly, so it has become a foremost example of prokaryotic gene regulation. It is often discussed in introductory molecular and cellular biology classes at universities for this reason" (https://en.wikipedia.org/wiki/Lac_operon, accessed May 12, 2016).

11. See https://www.genome.gov/11006943/human-genome-project-completion-frequently-asked-questions.

12. Baedke (2013, 770–71) provides an impressive taxonomy of the ways the epigenetic landscape is used but which stops, without comment, before it reaches the arts and humanities.

13. Keller (2000, S72) also suggests, in a fascinating footnote, that different conceptions of theory held by biologists, physicists, and mathematicians may explain why the "efforts to build a 'Theoretical Biology' that might productively interact with experimental molecular biology . . . have been notoriously unsuccessful." She seems to be speaking of the Serbelloni Symposia that Waddington organized in the late 1960s.

14. Richardson reports that even when an epigenetic study of sexual differentiation in rat brains (Nugent et al. 2015) produced results that seemed to affirm gender plasticity and phenotypic diversity, the finding of the study was instead reframed by its authors and by the scientific community as confirming "the paradigm of hardwired and dimorphic sex differences in the brain" (Richardson, forthcoming).

15. The epigenetic, even Waddingtonian, tropes in Mukherjee's understanding of cancer are even more apparent in his later "The Improvisational Oncologist." An extended landscape metaphor pervades this piece, from Christiana Couciero's illustration figuring an Ansel Adams photograph of a mountain range displayed on what seems to be a gridded guide for radiation therapy and pierced by an orange beam, to this extended analogy: "Michael Lerner, a writer who worked with cancer patients, once likened the experience of being diagnosed with cancer to being parachuted out of a plane without a map or compass; now it is the oncologist who feels parachuted onto a strange landscape, with no idea which way to go" (Mukherjee 2016a).

16. We return to the nature-versus-nurture controversy later in this study when we consider Keller's use of cartoons to explain that the allusion of a difference (or space) between the two concepts is a "mirage."

17. Waddington also considered, and discarded, another image of the epigenetic landscape: the phase space image. He later reformulated it as an "attractor surface" (Waddington 1957, 112). "Waddington was a serial modeler, a strategy which supported his approach to theory," Matthew Allen (2015, 119–42) has quipped. Over the years I have been writing this book, I have checked in with Clare Button, project archivist for the Towards Dolly: Edinburgh, Roslin and the Birth of Modern Genetics project at the Centre for Research Collections, University of Edinburgh, to ask about the artists and illustrators responsible for the second, third, and fourth versions of the epigenetic landscape, informally known as the "ball on the hill," the "phase space landscape," and the "view from underneath." Each time she has been unable to report any information. On May 11, 2016, after explaining, "I have been asked this question a number of times by researchers, but I'm afraid I have not so far been able to trace the name of the artist," she added, "If you do manage to solve the mystery, I'd be very interested to hear about it."

18. As he describes it later in the volume, the image seems a hybrid territory somewhere between scientific abstraction and fantasy. "It would be difficult to find a similar configuration in any actual piece of country. As one goes downhill, the valley which was originally wide and gently sloping, branches into more and more subdivisions, some of which (representing tracks realized only under the influence of special genes or environmental conditions) may be hanging valleys whose floors disembouch up the side, above the main valley bottom" (Waddington 1940, 93).

19. All three principles testify to Waddington's conception of epigenetics as a diachronic or developmental biology that could link embryology, genetics and evolutionary theory. He opposed this to a synchronic biology (physiology) that explored not progressive processes but cyclical ones (Gilbert 2000, 729; Waddington 1940, 2).

20. The phrase is Eric Lott's. I reflect on the privilege of being able to take these risky positions in Squier (2011, 14–15).

21. There is, of course, a much longer history to the concept of transversality. Lisa Diedrich (2016, 67) marks "the practice of transversality" as a "queer method," tracing it back to Félix Guattari's incorporation of it in his clinical work "as a theory and practice of ranging across identities, disciplines, concepts, and milieus in order to keep open the possibility of desire." The plenitude of intersections between Waddington's work and that of Gilles Deleuze and Félix Guattari is not something I cover in this book, alas, but it is a rich area for exploration.

22. I regret in particular that my approach only glances on the following crucial areas. By drawing on the epigenetic landscape as "a new tool or trope," researchers can study what Rob Nixon (2006–2007, 14) calls "slow violence," giving "symbolic shape and plot to formless threats whose fatal repercussions are dispersed across space and time." They can also recover a holistic vision of development that suggests alternatives to such slow violence (Fortun 2015, 36). Similarly, incorporation of images of the epigenetic landscape plays an increasing part in feminist scholarship on reproductive biomedicine, the politics of health and metabolism, and environmental and ecological issues studied through feminist new materialism and theories of the anthropocene (Alaimo and Heckman 2008; Davis and Turpin 2015; Landecker 2011; Lie and Lykke, forthcoming).

23. Tatjana Buklijas and Nick Hopwood, "Making Visible Embryos," online exhibition, 2008–2010, http://www.hps.cam.ac.uk/visibleembryos.

Chapter 1. The Epigenetic Landscape

1. The concept of canalization is subject to lively reconceptualization and discussion in molecular biology, genetics, and evolutionary biology even as I write: see http://en.wikipedia.org/wiki/Canalisation_%28genetics%29.

2. I am grateful to Ohad Parnes for sharing with me the English translation of his essay "Die Topographie der Vererbung" (2007).

3. Along with his wife, Myfanwy, he published *Axis*, a magazine of abstract art, which included in its November 1935 issue an article on Piper's work by Hugh Gordon Porteous titled "Piper and Abstract Possibilities" (Spalding 2009, 76).

4. The impact of that work registered in his writing about epigenetics as late as 1957, when, in *The Strategy of the Genes*, he introduced the concept of a creode: "The path followed by a homing missile, which finds its way to a stationary target, is a creode (Waddington 1957, loc. 347). Waddington would write about his war experiences many years later, in *Operational Research in World War II: O.R. against the U-boat* (1973).

5. As Waddington (1942, 563–65) defined the term, it referred to the process by which "developmental reactions, as they occur in organisms submitted to natural selection . . . are adjusted so as to bring about one definite end-result regardless of minor variations in conditions during the course of the reaction."

6. The participants were the molecular biologist Robin E. Monro; the theoretical biologist Brian Goodwin; the neuroscientists Jack Cowan, Richard L. Gregory, and Karl Kornacker; the geneticist John Maynard Smith; the theoretical physicist David Bohm, and the physicists E. W. Bastin, W. M. Elsasser, Martin A. Garstens, Edward H. Kerner, Paul Lieber, and Howard H. Pattee; the theoretical chemist Christopher Longuet-Higgins; the chemical engineer L. E. (Skip) Scriven; the systems analyst A. S. (Art) Iberall; the philosopher Marjorie Grene; and the automata theorist Michael A. Arbib. There was also a secretary, Miss D. Manning (Waddington 1969, 339–40).

7. Indeed, the disciplinary tensions in the conference are captured from the beginning of *Sketching Theoretical Biology*, the compilation of papers from the second symposium. A. S. Iberall, a systems analyst from Pennsylvania who took detailed notes throughout the event, characterized the experience as one of disciplinary disagreement in his personal overview. While the physicists seemed to have solved the problem of whether there could be a theoretical biology akin to theoretical physics by deciding in the negative (with some self-satisfaction), the biologists felt quite differently. For them, the central problem was the question of the ordering of biological development in time and space—or, as he imagined Waddington putting the question, "You physical fellows have had your fun, and I've paid for it, now where is the answer to my question? What is the theoretical basis for evolution?'" (Waddington 1967, 15).

8. I am referring to the published version of Waddington's paper; clearly he had revised it for publication, taking the opportunity to respond to the hostility to Bohm's defense of metaphysics. All quotations in this paragraph are from Waddington 1969, 72.

9. These concepts came to him during a tutorial in chemistry in his schoolboy days designed to prepare him for the university entrance examination, he tells his audience. His tutor, E. M. Holmyard, was obsessed with how Greek philosophy and technology were transmitted across space and time, from

the Alexandrian gnostics and the Arabic alchemists to fourteenth-century Europe. Holmyard branched into more esoteric studies, giving Waddington lessons in Arabic and exposing him to "a large number of very odd late Hellenistic documents" (Waddington 1969, 73).

10. When asked during an interview in March 2005, "What do you think about people who see you as a role model, as a pioneering professional female philosopher?" Grene responded, "I think they're pretty silly. They have no idea how different things used to be" (Cohen 2005).

11. The Irish elk exemplifies a scientific theory that is no longer accepted. According to the University of California's Museum of Paleontology, "The Irish elk was once considered a prime example of orthogenesis: it was thought that its lineage had started evolving on an irreversible trajectory towards larger and larger antlers. The Irish elk finally went extinct when the antlers became so large that the animals could no longer hold up their heads, or got entangled in the trees (http://www.ucmp.berkeley.edu/mammal/artio/irishelk.html, accessed August 13, 2016).

12. Grene also accused the symposium discussion of "incoherence" grounded "not just in mind-body dualism, which has long since given way to a belief in matter-in-motion as the sole reality," but also in the "deep-lying divisiveness of our conceptual framework" (Grene 1969, 62).

13. "I know that the fashion of the time is very much against notions which have such transcendental-sounding names as Whitehead's 'eternal objects'; but I shall require convincing that David Bohm's 'timeless orders' are either very different, or much more likely to prove acceptable to the philosophical establishment" (Waddington 1969, 69).

14. As Grene (1995, 130) points out, Dreyfus was interested in the former and she in the latter questions.

15. "Such a catastrophe (i.e., a shift in the landscape's qualitative form by coalescence of a peak with a valley, resulting in the annihilation of the later [sic]) can be visualized as a 'ball' on a surface jumping across regions where no equilibrium points are defined to a new stable 'equilibrium region' (i.e. a different valley)" (Baedke 2013, 762).

16. In a later chapter we will see this disciplinary gatekeeping fade away, first within medicine, and later in communication between medicine and the lay public. For example, extrapolating from relativity theory in physics to what he calls "biological relativity" to model cardiac behavior, the physiologist Denis Noble (2011, 55) argues that there is "no privileged level of causation." Rather, "Biological relativity can be seen as an extension of the relativity principle by avoiding the assumption that there is a privileged scale at which biological functions are determined."

17. "Ventriloquism," Dictionary.com, http://www.dictionary.com/browse /ventriloquism?s=t, accessed April 9, 2015.

Chapter 2. A New Landscape of Thought

1. I am using the concept of canalization as a heuristic, a trope that may help illuminate aspects submerged in the accepted narrative of Waddington's work at the Serbelloni Symposia. Of course, if our focus is Waddington's own intellectual development, the symposia might instead be thought of as an example of failed canalization, in which the gendered and disciplined influence of his environment interrupted, redirected, and squelched his emerging holistic, interconnected understanding of development to a reductive determinism.

2. See http://rationallyspeaking.blogspot.com/2011/07/meaning-of-theory-in-biology.html. Unlike Pigliucci's discussion here, we will see that contemporary uses of the epigenetic landscape can have a heuristic function (see Baedke 2013).

3. I thank Bruce Clarke for this observation.

4. I discuss these gender dynamics at length in an earlier version of this essay (see Squier 2013).

5. Susan Oyama (2015) discusses just this problematic relation between "chance" and "determinism" in developmental systems theory's concept of "contingency."

6. The ever useful Wikipedia defines a phase space as "a space in which all possible states of a system are represented, with each possible state of the system corresponding to one unique point in the phase space" (http://en.wikipedia.org/wiki/Phase_space, accessed June 4, 2015). Google's definition, accessed on the same day, is more precise as to the origins in physics: "a multidimensional space in which each axis corresponds to one of the coordinates required to specify the state of a physical system, all the coordinates being thus represented so that a point in the space corresponds to a state of the system" https://www.google.com/webhp?sourceid=chrome-instant&ion=1&espv=2&ie=UTF-8#q=phase+space&*. Waddington's definition was somewhat longer: "A system containing many components can be represented in multidimensional space, the co-ordinates of the point in each dimension representing the measure of a particular component. . . . As the composition of the system changes the point will move along a certain trajectory" (Waddington 1957, loc. 561).

Chapter 3. Embryo

1. I thank Vanessa Lux for reminding me of this crucial point when I presented this material at the Zentrum für Literatur- und Kulturforschung, Berlin.

2. The verses are in part:

Happy the egg, the bubble blastula,
Does nothing much, but all the same,
Each part is as good as every other
Or would be if it could
But better its situation;

The opportunist roof, the sodden floor
With a space between:
A capital notion, but a lazy affair.

Feeling, thinking, feeding come
By limiting this freedom
Of each being any
Now each for all
Performs its special function.
Now every separate part is tied
To particular performance
But still within itself is free
To organize its own affair.
(Squier 2004, 79–80)

3. "Embryos are inert cells; they do not grow, or undergo any significant gene expression, or have any capacity to develop further until they are implanted into a gestating uterus" (Maienschein 2014, 280).
4. In creating the second version of the epigenetic landscape, Waddington employed traditions of scientific imaging at some distance from the landscape drawing that informed Piper's river scene, borrowing from the cell biologists and population biologists before him who had used versions of a landscape to represent development.
5. Throughout this section I rely on "Making Visible Embryos," an online exhibition created by the historians of science Tatjana Buklijas and Nick Hopwood and hosted by the University of Cambridge and the Wellcome Institute. It offers a valuable schema of the visible embryo from the fourteenth century to the 2000s (Tatjana Buklijas and Nick Hopwood, "Making Visible Embryos," online exhibition, 2008–2010, http://www.hps.cam.ac.uk/visibleembryos).
6. Of course, this religious view influenced even medical images of the period. Girolamo Mercurio's anatomical image of a pregnant woman lifting the veil of her belly to show the unborn "creature" inside reflected the conventions of religious iconography: head up, eyes once again locked with the viewer's. In a remarkable and improbable return of that theological vision, one contemporary right-to-life website, Unbornwordoftheday.com, offers images that claim to harmonize the theological and the scientific, showing images of the Zygote Christ Child, the Blastocyst Christ Child, the Embryonic Christ Child, and the Fetus Christ Child (http://unbornwordoftheday.com/2010/12/05 /what-does-my-little-savior-look-like-first-trimester).
7. Buklijas and Hopwood, "Making Visible Embryos."
8. As we see at the end of this chapter, Wellmann disagrees with this position, arguing that Sömmerring's reliance on size as the principle of sequence made him unable to take a position in the preformation-epigenesis debate: "Soemmerring found himself unable to champion either theory, and instead

declared himself undecided. In his essay on deformities, he wrote that he could 'avow neither of the existing series of generation.' Both the theory of reformed germs and that of epigenetic development 'contain, in my opinion, truths that may very well and easily be combined with the truths of the other; yet neither of them alone seems to me exclusively true and satisfactory'" (Wellmann 2017, 191–92).

9. Buklijas and Hopwood, "Making Visible Embryos."
10. Weiss delivered these remarks, according to Gilbert and Faber (1996), to a meeting of the American Association for the Advancement of Science in 1955, opening with, "Beauty is order; life is order; hence life is beauty. It is a syllogism—that simple."
11. Buklijas and Hopwood, "Making Visible Embryos."
12. Buklijas and Hopwood, "Making Visible Embryos."
13. In an insight discussed at greater length in what follows, Hopwood (2006, 260, 298–99) has argued that the controversy over Haeckel's embryo images should move us to "take seriously how meanings are made as pictures move between classroom teaching, research programs, and public debate."
14. As Hopwood (1999, 464) points out, his use of three-dimensional modeling may cause us to "revise our accounts of scientific controversy and change." "Precisely because they have been so disputed, bringing models in three dimensions into general histories of the sciences may do more than just fill them out. By studying models that some groups promoted or relied on and others ignored or opposed, we can develop new views of what was at stake in struggles over change."
15. "For this seriation, His used the term *Normentafel*, or "plate of norms." It would later either be kept in the German or translated as "normal plate" (His 1885; Hopwood 2000, 38). There is some slippage in meaning here: norms were standards, but for human embryos recovered from abortions, the main concern was to avoid abnormality" (Hopwood 2007, 4).
16. It is important to acknowledge here that the distinction between "embryo" and "fetus" is an imprecise one, with both terms existing as long ago as the fourteenth century and divided by a shifting and culturally inflected line. As Lynn Morgan points out, "Historically, the distinction between 'embryo' and 'fetus' corresponded to the distinction between organisms that did and did not look human, that is, that had human form. . . . *Embryo* is the term traditionally given to organisms that have not yet achieved that form, while *fetus* refers to 'the unborn young of any vertebrate animal, particularly of a mammal, after it has attained the basic form and structure typical of its kind'" (*Encyclopedia Britannica* 2003; Morgan 2009, 92). "What we refer to as 'embryo' is not a stable ontological thing, but a recent, tenuous, and ever-shifting social consensus about the meanings we are willing (though not without controversy) to ascribe to certain physiological properties. In the midst of this sematic contestation, people negotiate their power—and the ontological

status of embryos—by appropriating terminology to suit their own goals" (Morgan 2009, 94).

17. Hopwood (1999, 463–64) has argued that we do well to study such three-dimensional scientific models not only to "deepen our appreciation of the variety of representational work in the sciences," but also because "by studying models that some groups promoted or relied on and others ignored or opposed, we can develop new views of what was at stake in struggles over change."

18. Buklijas and Hopwood, "Making Visible Embryos."

19. Later chapters show that the contemporary landscape architects Corner, Mathur, and da Cunha, like many contemporary cartoonists, view drawing as a crucial mode of thinking.

20. Needham was one of the founding members, along with Waddington and the mathematician Dorothy Wrinch, of the Theoretical Biology Club, a study group that devoted itself to solving "The Great Problem": "What is the relation between those large particles which we call elephants, trees, or men, and those extremely small ones which we call molecules or electrons?" (https://jennymcphee.wordpress.com/tag/the-theoretical-biology-club; see also http://www.bookslut.com/the_bombshell/2013_03_019986.php, accessed May 2, 2015).

21. The spelling Yūgen comes from the Stanford Encyclopedia of Philosophy entry on Japanese aesthetics. https://plato.stanford.edu/entries/japanese-aesthetics/. In the German embryological tradition it is Yügen.

22. The influence of Yūgen and Kire shaped the work of the contemporaneous filmmaker Yasujiru Ozu, whose films drew on both. The British film critic Basil Wright explains how these films used cut and continuity to capture a specifically emergent perspective: "the films and the people in them unfold before ore [sic] us like those paper flowers—also Japanese—which open gently to reveal their petals when we drop them into a bowl of water" (quoted in Crittenden 2003, 21).

23. Yet with the embryo collector's respect for the individual specimen, Streeter at first fought shy of "the term stage, with its implication of precision." Instead, he segregated embryos more flexibly into "age groups" that "represent levels in their structural organization as a whole." He borrowed the term "horizons" from geology and sought, like fossils for strata, several morphological criteria for each one (Streeter 1942, 213–14; see also Hopwood 2007).

24. At times, these images were verbal or tonal, as in the case of Goethe's treatise and poem of 1790 and 1798, both titled *The Metamorphosis of Plants*, and the work of such eighteenth-century music theorists as Johann Georg Sulzer, for whom the essential principle of musical beauty was "a state of tension" (Wellmann 2017, 68).

25. "The emergence of the rhythmic episteme in the period around 1800 was not the transfer of an aesthetic perspective onto natural history. . . . Neither

was the poetological [*sic*] and philosophical discussion of rhythm merely a 'reception' of scientific concepts or vice versa. Rather, the idea of rhythmically organized nature rested on an episteme of rhythm that *simultaneously* formed the foundation of new aesthetic concepts in literary and music theory and was articulated in scientific theories—whether in the epigenetic theory of generation, Goethe's model of the metamorphosis of plants, or the growing currency of the physiological concept of the alternation or transformation of matter, in the contemporary German discourse, *Wechsel der Materie*" (Wellmann 2017, 19).

Chapter 4. The Graphic Embryo

1. Indeed, we could bring their narrative still farther along its track by exploring the use of algorithms in Hugo de Gris's EMBRYO project, which in 1991 envisioned the new field of "artificial embryology called embryonics (short term for 'embryological electronics'). Embryonics is based on the idea of adapting the processes found in embryonic development to build artificial systems." (Julia Damerow, "'Genetic Programming: Artificial Nervous Systems, Artificial Embryos and Embryological Electronics [1991], by Hugo de Garis," in *Embryo Project Encyclopedia*, June 10, 2010, http://embryo.asu.edu/handle/10776/2012.

2. "Steamboat Willie," *Variety*, November 21, 1928, Variety Talking Archive, 1905–2000, 93.6.

3. "Comics Unmasked: Art and Anarchy in the U.K.," *British Comics Collection*, http://www.bl.uk/reshelp/findhelprestype/news/britcomics/#further, accessed February 27, 2015.

4. "Comics Unmasked"; "Alfred Harmsworth, Lord Northcliffe," *Oxford Reference*, http://www.oxfordreference.com/view/10.1093/oi/authority .20110803095921726, accessed February 27, 2015.

5. As this book was going to press, I discovered with pleasure Richard Grusin's article "Radical Mediation" (2015), in which he draws some of the same connections between Whitehead's notions of concrescence and prehension, Simondon's notion of transindividuation, and the relations between biological and cultural developments that I have been tracing in this volume.

6. As we see in the next chapter, in an examination of the work of the landscape theorists Mathur and da Cunha, the act of mapping can falsely stabilize a river that is actually flexible, shifting, and even agential. Thus, a reimagined ontology of water generates promising new ways of working with the river in landscape design, rather than persisting in doomed attempts to manage the river or contain flooding. This process resembles the effects of the reimagination of human development made possible by graphic embryos.

7. See Ian Williams, "Why Graphic Medicine"? http://www.graphicmedicine.org /why-graphic-medicine, accessed June 9, 2015.

8. "The term 'evidence-based medicine,' as it is currently used, has two main tributaries. Chronologically, the first is the insistence on explicit evaluation of

evidence of effectiveness when issuing clinical practice guidelines and other population-level policies. The second is the introduction of epidemiological methods into medical education and individual patient-level decision-making" ("Evidence-based Medicine," Wikipedia, http://en.wikipedia.org/wiki/Evidence-based_medicine, accessed March 27, 2015).

9. Tatjana Buklijas and Nick Hopwood, "Making Visible Embryos," http://www.hps.cam.ac.uk/visibleembryos; see also Hopwood 2015.

10. The use of rhythmic pictorial series to depict embryonic development has been attributed to several influences. The historian of science Janina Wellmann (2017) links them to a new type of pamphlet and book that appeared in the 1800s, "instructional graphics" that used a series of sequential illustrations for instruction in the art of scripted movement, whether in the art of military drill, gymnastics, or dancing. Wellmann argues that they "supplied the basic iconographical framework for one of the most important forms of visualization in the new field of biology: the representation of embryonic development as a picture sequence."

11. Encouraging further exploration of "the production, manipulation, and display of three-dimensional representation devices in embryology and in the sciences, technology, and medicine more generally," Hopwood (1999, 496) encourages us to resist our tendency to "fetishiz[e] an isolated class of objects," but rather to ask what these innovations in embryological modeling can tell us about "changes in making other medical or biological, geological, engineering, chemical, mathematical, or physical models." My turn to graphic embryos is in part a response to that call.

12. I use the word consumes not in the sense of purchasing but in the sense of ingesting, taking in, incorporating as well as being rendered corporeal by the comic.

13. Phenotype, from the Greek *phainein*, meaning "to show" and *typos*, or type, refers to all of the observable aspects of an organism: its shape, structure (inside and out), biochemical or physiological composition, as well as its traits, behaviors, and even the products of its behavior. In postgenomic medicine, this focus on genotype and phenotype has gradually been replaced by an attention to what Keller has called the "vast reactive system than as a collection of individual agents . . . that bear responsibility for phenotypic development" (Keller 2015, 10).

14. I thank the science studies scholar and cartoonist Jenell Johnson for reminding me of this nuance.

15. Some of the discussions in this section appear in rather different form in my co-authored book *Graphic Medicine Manifesto* (Czerwiec et al. 2015). Although I am aware that contemporary publishing protocols discourage such repetitions, in my view thinking proceeds much as do all other (scaled) forms of development: through rhythmic repetition with a difference. Over the years I have been thinking and rethinking these interesting comics, and in the

process my readings of them have (much like life) changed and become more complicated.

16. Buklijas and Hopwood, "Making Visible Embryos." See also Morgan, "A Social Biography of Carnegie Embryo No. 836."

17. From Joyce Farmer's *Abortion Eve* on, these comics have had a powerful feminist impact, registered in Hillary Chute's *Graphic Women*. See also Squier 2015.

18. As we will see in the next chapter, Nilsen's comic also recalls the ambitious vision of Jantsch and Waddington, *Human Consciousness in Transition*.

19. Anders Nilsen, "Me and the Universe," September 24, 2014, http://www.nytimes.com/interactive/2014/09/25/opinion/private-lives-me-and-the-universe.html.

20. Manfred Gödel, personal communication, August 25, 2014, Art Laboratory Berlin.

21. I am thinking in particular of Lynn Margulis's reconceptualization of evolution in her work on the microbial origins of human life through symbiogenesis and lateral gene transfer, and the work of James Shapiro on cell-cell signaling (Margulis and Sagan 2002; Shapiro 2015).

22. Recent challenges to the concept of the Anthropocene have argued that it reflects the perspective of the privileged, consumer-oriented North, rather than a global perspective. Donna Haraway and others are suggesting that a better term would be "capitalocene," while Louise Brown, in a forthcoming study, argues that the Anthropocene neglects the post-colony perspective of Africa. For a glimpse of Haraway's recent argument, see Haven 2014 and the accompanying video at https://vimeo.com/97663518; Louise Brown, *The Nature Industry*, unpublished ms., courtesy of the author.

23. In a later chapter we will see that there is a similarity between Nilsen's "Me and the Universe" comic and Ian McHarg's incorporation of the methods of epidemiology into landscape architecture. But there is a crucial difference, too. Unlike McHarg, whose admiration for epidemiology testifies to his preference for the scale of the population, Nilsen deliberately works back and forth between the population and the individual, embracing and asserting the connection between scales while maintaining the important difference in perspective they provide. In this way, his approach to the embryo echoes what we will see when we examine Mathur and da Cunha's approach to the landscape; for all of them, remediation necessarily involves both the individual and the population, both the embryo and its landscape.

24. Ian Williams, "Sick Notes," http://www.theguardian.com/lifeandstyle/series/sick-notes.

25. In its incorporation of Tarot card imagery, *The Bad Doctor* (Williams 2014) pays homage to *Binky Brown Meets the Holy Virgin Mary* (San Francisco: Last Gasp Eco Funnies), Justin Green's pathbreaking narrative of OCD.

26. The connection between twins and worries about pregnancy outcome is understandable, but epigenetic medicine has recently escalated those worries, finding genetically identical twins fertile research soil for examining epigenetic factors

in psychiatric illness. "Twin Study Reveals Epigenetic Alterations of Psychiatric Disorders," press release, Kings College London, http://www.kcl.ac.uk/ioppn/news/records/2011/Sept2011/epigeneticsofpsychiatricdisorders.aspx, accessed May 3, 2015.

27. I have searched in vain for the blog on which I found a mention of the influence of *Tango* on McGuire's comic. To that insightful blogger, my apologies. For some of the many illuminating readings of this comic, see: http://www.nytimes.com/2014/12/24/books/here-richard-mcguires-new-graphic-novel.html?_r=0, http://boingboing.net/2014/12/11/here-is-richard-mcguires-epi.html, http://www.theguardian.com/books/2014/dec/17/chris-ware-here-richard-mcguire-review-graphic-novel, http://www.theatlantic.com/entertainment/archive/2014/09/richard-mcguires-time-machine-with-a-view/380736, http://penguinrandomhouse.ca/hazlitt/blog/eternal-returns-american-room, http://harpers.org/blog/2014/12/time-out-of-joint-in-richard-mcguires-here, and https://www.bostonglobe.com/arts/books/2014/12/27/book-review-here-richard-mcguire/v36ezrasoFfpeXmCGKISwJ/story.html.

28. See http://www.zbigvision.com/Tango.html; Rybczynski 1997.

29. "Bouncing Ball," Wikipedia, http://en.wikipedia.org/wiki/Bouncing_ball.

30. The film adaptation of "Here," made by Timothy Masick and William Trainor in 1991, makes it clear that the birth of a baby is the triggering event in this spatial/temporal mashup. Ed Howard, post to Only the Cinema blog, February 22, 2008, http://seul-le-cinema.blogspot.com/2008/02/222-here.html.

31. *Tango*, https://www.google.com/webhp?sourceid=chrome-instant&ion=1&espv=2&ie=UTF-8#q=tango+film+zbigniewa+rybczy%C5%84skiego&*. Accessed February 28, 2017.

32. Catherine Malabou's (2006, 431) trenchant observation that contemporary philosophy "bears the marks of a primacy of symbolic life over biological life that has been neither criticized nor deconstructed" testifies to the fact that the scientific side of this interaction is not the only place where there are such disciplinarily primed resistances.

33. See http://www.epigenesys.eu/en/in-the-news/all-news/1304-hashtag-visions-of-epigenetics.

Chapter 5. The River in the Landscape

1. Both born in the highlands of Scotland, Waddington and McHarg also shared space in the pages of the *Daily Pennsylvanian* on March 17, 1970, when their paths converged briefly in Philadelphia. Waddington, then Einstein Professor at the State University of New York, Buffalo, gave a lecture titled "Developmental Organization Purpose and Ethics" at the University of Pennsylvania Medical School; the same morning, McHarg had held a press conference in the university's Fine Arts Gallery to announce plans to launch the first Earth Week. That brief spatial proximity in 1970 seems fitting, for it was at the University of Pennsylvania that McHarg transformed the field of landscape

architecture by incorporating the insights of ecology—insights that Waddington would soon include in the collection *Evolution and Consciousness*, co-edited with Erich Jantsch, in the form of an essay contributed by the Canadian ecologist C. S. Holling.

2. That Serres was influenced by Waddington seems likely. Certainly, Serres makes the explicit distinction between his concept of "homeorrhesis" (with its thermodynamic framing) and that of "homeorhesis," which he places at the intersection of ontogenesis and evolution: "Let us shift from the global to the local level, from the whole of the organism to the diverse systems that used to be called respiratory, circulatory, neurovegetal, and so forth, and then to organs, tissues, cells, molecules. . . . The passage could be plotted from homeorrhesis to homeorhesis" (Serres et al. 1982, 76).

3. In a lovely example of the widespread engagement in the epistemological value of rhythmic embodiment, McHarg shares this view not only with the other landscape theorists I discuss but with his predecessor embryologist Wilhelm His and the cartoonists involved in the Graphic Medicine movement.

4. For a powerful example of feminist ethical engagement with unfolding biological development on multiple timescales, see Murphy 2013. See also Margulis et al. 2002.

5. As I show in chapter 6, Meyer's recent work explicitly lays claim to feminism as a core principle of an expanded field of landscape architecture (Meyer 2011).

6. Technics is a "structure of inheritance and transmission [that] is external and non-biological," Andrès Vaccari and Belinda Barnet (2009) explain. "It is not genetic but acquired. It is in this sense that Stiegler calls the structure 'epigenetic': it exists outside and in addition to the genetic. . . . Stiegler expands on the current scientific definition of epigenetic processes to include this cultural dimension, but he also demarcates a third layer, a structure that stores and accumulates individual ontogenesis, and which exists beyond the central nervous system, beyond genetic memory. This contains both culture (as collective knowledge) and technical artefacts. It is a place of storage, accumulation and sedimentation of successive events, a thing that evolves and has its own dynamic, its own history and historicity. . . . Stiegler calls this the *epi-phylogenetic* level, implying by that terminology a material genealogy proper to it. Expanding on Leroi-Gourhan, he distinguishes here between three types of memories out of which the human (at both the individual and species level) develops: genetic memory; memory of the central nervous system (epigenetic); and techno-logical memory [epiphylogenetic]."

Drawing on Simondon, Richard Grusin (2015, 142) proposes the notion of "radical mediation" as "another way to talk about what Whitehead calls occasions of experience, or the fact that life or nature always involves duration and persistence and movement." He goes on to make "my stronger claim . . . that these activities of radical mediation (remediation and premediation) constitute the ontological character of the world . . . similar to how Whitehead uses occasions of experience, process, or event to characterize the world as in a constant state of creation and re-creation."

7. Christian de Lutz and Regine Rapp situate their work in relation to Eliasson's, emphasizing his interest in creating a spectacle rather than collaboratively engaging with his environment. (Interview with the author, 19 May 2015.) However, "Riverbed 2014," Eliasson's installation at Denmark's Louisiana Museum of Modern Art, seems to me to engage in a beautifully extended remediation of the Piper drawing (Olafur Eliasson, "Riverbed 2014," http://olafureliasson.net/archive/artwork/WEK108986/riverbed#slideshow.

8. "Programmed" is the term used in landscape architecture for designing ways a landscape will be moved through, experienced, and used by its inhabitants or visitors.

9. Note, too, the parallel to Waddington's interest in Sewell Wright's "fitness surface" as a model for representing developmental change, before he settled on his own image of the epigenetic landscape (Waddington 1957, loc. 1997).

10. I draw partly on Anne Whitson Spirn's description and analysis of this project in the discussion that follows, as well as on my own reading of the Pardisan plan, written by McHarg and his associates W. Robinson Fisher, Anne Spirn, and Narendra Junega (Mandala Collaborative 1975).

11. We will come across that geek again in chapter 7, where she or he appears as the preferred visitor at Art Laboratory Berlin. See also Sagan 2007, 2012.

12. This recourse to aesthetics other than Western ones to guide thinking about the relation between the individual and the environment not only recalls the inspiration Waddington drew from the Alexandrian gnostics and the Arab alchemists, but it also connects (down the scale, if you will) to an aspect of the embryological aesthetic, as I show in chapter 3.

13. See http://www.charlesjencks.com/#!architecture-and-sculpture.

14. See https://www.jupiterartland.org/artwork/metaphysical-landscapes; https://www.jupiterartland.org/artwork/cells-of-life.

15. See http://www.charlesjencks.com, accessed April 21, 2015.

16. The phrase echoes the description of the Liesegang rocks on Jencks's website, from which Jencks has formed concrete seats in the shapes of cell models that "bear an uncanny relationship to the many organelles inside the units of life" (https://www.jupiterartland.org/artwork/cells-of-life).

17. Charles Jencks, interview with the author, March 20, 2015, London. To offer one example of the exciting turbulence of our conversation, which circled from metaphysics to molecular biology, consider Jencks's description of how he consulted the molecular embryologist Marilyn Monk to get her advice on constructing an epigenetic landscape. It was Monk and her group who discovered the phenomenon that she named "deprogramming," by which epigenetic programming of the parental genes is erased, returning the embryonic cells to their initial, totipotent state: "So I'm realizing that there're many ways to describe the same thing, which means that evolution must, as Darwin himself

said, have many flavors. OK, and there must be more than one mechanism, and how do you describe the general area, because if you say it's the epigenetic revolution or the epigenome as a whole, that's controlling, [that] tells the genes what to do, which I'm prepared to say [is] a bit like hand waving. Fine, you've told us the epigenome does the work, but who tells it how to work? And then Marilyn Monk says, 'Well, it's the metabolome . . . who tells the metabolome—'What is really going on here? Who's marking all these things?' So I think I've retreated to saying, 'How am I supposed to know?' I'm not even in the field, and it looks to me like there're four separate levels of description, of which evo-devo is one [and] genetic simulation overlaps—perhaps it's the same thing. The Hox gene is another one. The RNA world is another one, and in some way each one of these flavors of evolution, of evolutionary mechanism, are external factors in evolution, . . . it's like a black box. I come back to the black box. We don't know what is inside the black box, but there is a black box, and these different things are happening, all of which have been described by different parts of biology. Explain why evolution is so fast? It's incredibly fast, actually. Much faster than simple vanilla-flavored Darwinists can explain, and therefore there's something crazy about our synthesis . . . And [Arthur] Koestler was on to this: it's rather like the history of cosmology" (Jencks interview).

18. See http://www.charlesjencks.com/#!the-garden-of-cosmic-speculation.
19. See www.charlesjencks.com/#maggie's-centres, accessed April 20, 2015. During our interview, Jencks confirmed that his Maggie's Centre work connected to epigenetic medicine and referred to Norman Doidge's work on neuroplasticity: "*The Healing Brain*, he'd just come out with it. He was just here. I heard him speak—sell-out audience at King's Place. . . . Doidge's thing is neuroplasticity, the neuroplastic revolution. So I'm seeing, God, this is part of the same paradigm. . . . I'm thinking this is exactly what Maggie and I were talking about when we set up what was named after her [Maggie's Centres] that it matters. Like the epigenome is essential for Maggie's Centres. If you have cancer, you can still do things, even terminal cancer, even six months to live; you can still make it eight months. You can't beat it, because actually the cancer is living. The cancer is alive and will go around you often. That's a whole other thing, but I think that it's really key, this paradigm shift. It's what I call the postmodern" (Jencks interview). Jencks's work is uncanny, then, in such inherent contradictions as can be grasped if one juxtaposes his exhibitions of sculptures and landforms at an "ancient place with reputed Knights Templar connections" (that military order so powerfully connected to the purse of the Roman Catholic church) to his commitment to the nonprofit mission of the Maggie's Centres to provide "information and support that is unconditional and institutional, and therefore, all the more powerful and effective," according to the Scottish radiation oncologist Alastair Munro (https://www.maggiescentres.org/about-maggies).

Chapter 6. Designing Rivers

1. The Internet is replete with images of the Mississippi River Basin Model. A good general source is Wikipedia (https://en.wikipedia.org/wiki/Mississippi _River_Basin_Model).

2. Let me elaborate on the difference between the MRBM and its successor, the "layer-cake method" introduced by Ian McHarg some twenty years later. The MRBM attempted to model hydraulics in a homeostatic system, introducing different levels of water flow into a series of fixed and stable streambeds, rather than modeling true homeorhesis. While the waters would rise and fall, their channels were unchanging. In contrast, McHarg conceptualized a region as a layered "cake" of different processes, not different static entities. Everything in the region, from the granite substrate to the water table, the microbes, the flora and fauna, and the climate, McHarg thought of as *in flow*, in process, and it was those processes he juxtaposed or layered on top of one other to come up with his regional assessment. His additional planning document, the matrix, simply offered another way to grasp that assessment, subdividing or sectioning it not horizontally (if you will) but vertically instead. Thus, when we move from the MRBM of the 1940s to the layer-cake model of the 1960s, we see a decisive redirection in the flow of landscape thinking. However, this was a redirection, not an arrival. McHarg's method could not achieve homeostasis, could not stay useful across all conditions, as was evident in the fact that McHarg ultimately was unsuccessful in imposing it in the very different climatological and political context of the Pardisan project in Tehran. Instead, the firm accrued crippling debts in the firm with the fall of the shah of Iran and the failure of the Pardisan project, and McHarg was forced to abandon his professional work with the layer-cake model.

3. "The following model has been inspired by an idea given by C. H. Waddington, the idea of the epigenetic landscape: C. H. Waddington proposed to look at the development in embryology as the trajectory of a material point in a 'landscape' where valleys define the main paths of development. . . . We may materialize the model in three-space by shrinking the number of variables from three to two. We realize then a kind of potential well. . . . We then flood this well by pouring water in it. The shore of the lake so obtained describes the spatial development of the embryo; there are three main valleys in the geography of the potential well corresponding to the three main layers: ectoderm, mesoderm, endoderm. . . . Another interest of this model is to make possible a representation of (vegetative) reproduction. High above the mesoderm valley is a suspended lake symbolizing the gonad. . . . Suppose at the bottom of the gonad lake there is a small pipe pouring above the germinal point of M_2 (In fact, in mammals, the pipe connecting M_1 to M_2 has a kind of anatomical realization in the umbilical cord.) . . . Despite the obvious shortcomings of this model . . . it gives some reasonable intuition of the global dynamic of reproduction in the living beings" (Thom 1970, 112–14).

4. Sam Wallman, "So Below: A Comic about Land," 2016, http://sobelow.org, accessed August 6, 2016.

5. She continues, "The site descriptions of [Andreas] Duany and [Elizabeth] Plater-Zyberk are examples of this, as any landscape element that alters the town plan grid is considered awkward or distorting. There are not two structures on the site, only one—that of Euclidean geometry." For more on Duany Plater-Zyberk and Company, see http://www.dpz.com/Projects/All, accessed August 6, 2016.

6. Their co-edited collection, *Design in the Terrain of Water* (Mathur et al. 2014a, 1, 5) brings this theoretical discussion to a transdisciplinary group of people working in a number of different professions, asking them all to contemplate the political, economic, and social meaning of water. Tracing the lines that define rivers back to the mapmaker Anaximander of Miletus, they propose in their introduction that "the lines of rivers are not universal, but rather products of a particular literacy through which water is read, written, and drawn on the earth's surface, on paper, and in the imagination. Perhaps it is time to ask if this literacy has run its course; and . . . we need to take down the towering edifice of imagery, beliefs, policies and practices that have been built for over two millennia on river banks in order to make room for a terrain that can be engaged differently, such as with an appreciation of waters everywhere." See also Pevzner and Sen. 2010.

7. Dilip da Cunha, March 5, 2015. Author's interview with Anuradha Mathur and Dilip da Cunha via Skype, March 5, 2014.

8. Anuradha Mathur and Dilip da Cunha, interview with the author, Skype, March 5, 2015.

9. See http://www.crwa.org/charles-river-history, accessed April 27, 2015.

10. Mathur was trained at the Center for Environmental Planning and Technology in Ahmedabad; da Cunha was trained at the School of Planning and Architecture in New Delhi.

11. Dilip da Cunha speaking. Mathur and da Cunha interview, March 5, 2015.

12. Stewart is drawing, of course, on Eve Sedgwick's concept of "weak theory" (Sedgwick 1997; Stewart 2008). Although Stewart doesn't make the point, Sedgwick's weak theory anticipates the turn from critique to concern found in the philosophy of Michel Serres and his ardent follower, Bruno Latour.

13. da Cunha, in Mathur and da Cunha interview.

14. See also http://www.soak.in/exhibition_intro.html.

15. See http://structuresofcoastalresilience.org.

16. Full disclosure: my daughter, Caitlin Squier-Roper worked as the project manager and lead designer on the project. I thank her for introducing me to the visionary work and delightful presences of Mathur and da Cunha.

17. When I asked them in our interview in 2015, "Could you say something about the many ways you understand development," da Cunha responded, "That's such a loaded term. . . . We shun the linearity of development, on the one

hand. On the other hand, actually, we move more towards potential and initiation. So one of the things that we have been trying to do and to try and clarify actually; design as initiation rather than even process. How do you begin to release potential?" Mathur said, "You don't want control, but you've given it a certain tendency, given it a certain direction." Later in the same discussion, da Cunha added, "Development actually carries a universal intent that places like India are sort of jumping on. It's . . . a bandwagon, and you know, it carries—from corruption to everything else. But even if it's of good intentions, actually, it's the universal that takes over. And what we are concerned with, actually, is that if we had to engage the particular, then we are in a much better position to initiate . . . movement and place."

Chapter 7. "A Complex System of Interactions"

1. Mitchell's discussion of bioArt's incorporation of the work of Gilbert Simondon—particularly his principles of "metastability" and "individuation"—was important to my understanding of this concert. See the SymbioiticA website, at http://www.symbiotica.uwa.edu.au. As I would learn later, ALB had mounted previous shows embodying each of these modes of vitalist bioArt.

2. Ludwik Fleck had a phrase for this. In English, it is translated as "thought community." But that translation is inadequate. In German, the Fleck scholar Martina Schlünder tells me, it really reads more like "thinking community." The distinction is crucial, because it captures the theorist's interest in thinking as a communal process rather than the isolated product of one individual's mind (Martina Schlünder, personal communication).

3. As Silvia Caianiello (2009, 76) has argued, in choosing this visual image Waddington moved beyond the atomistic notion of evolutionary fitness. "In Waddington's view, fitness is namely a systemic property, 'a quality of the organism as a whole'" that cannot be accounted for by simply breaking it down "into a number of immediate components" (Waddington 1957, 110).

4. Caianiello elaborates on Waddington's reasons for rejecting earlier models of development, particularly Sewell Wright's adaptive landscapes and Ashby's phase-space model of the epigenetic landscape.

5. The other elements of this list are "parity of reasoning"; "the developmental and evolutionary interdependence of organism and environment"; "a shift from 'genes and environment' to a multiplicity of entities, influences, and environments"; "extending heredity"; "a shift from central control to interactive, distributed regulation"; and "a shift from transmission to continuous construction and transformation." Oyama's work not only echoes Waddington's stress on scale; it is also consonant with the approach of Bruno Latour in Oyama's insistence on symmetry in the exploration of developmental systems (Oyama 2000, loc. 177). According to Oyama and her colleagues (2002, 1), DST offers us the "stake-in-the-heart move" by which we can escape these confounding and reductive either-or traps.

6. "Moving among scales in this way not only enables us to see more, it also gives us a more acute sense of the many relations among these scales," Oyama argues.

7. I thank Irina Aristarkhova for this illuminating question.

8. This description of the multiscale nature of the biological picture provides an opening to consider the relations among different disciplines as they intervene at different scales. Eva Jabloncka, Marion Lamb, and Anna Zeligowski integrate multiple disciplines into their analysis of epigenetics, arguing that a scale shift "is now being incorporated into an evolutionary-developmental (evo-devo) framework, and also into ecological and medical research" (Jablonka et al. 2014, 454).

9. Material in quotation marks comes from the ALB website at www.art laboratory -berlin.org/html/eng-team.htm.

10. I borrow this term from Melinda Bonnie Fagan (2012, 189), who has argued that in its unification of genetics, development, and evolution, this image of the epigenetic landscape constituted a "conceptual laboratory" for Waddington.

11. See http://www.creative-city-berlin.de/en, accessed April 26, 2015.

12. "Die Kolonie versteht sich als Möglichkeitsraum, dessen Projekte, Initiativen und Kooperationen sich nicht auf Vermarktungskriterien beziehen. Die Heterogenität ihrer Mitglieder und der gezeigten Kunst zeichnen die Arbeit des Kunstverbundes aus. In selbstorganisierten Ausstellungen finden sich Auseinandersetzungen mit politischen, gesellschaftlichen und interkulturellen Themen in künstlerisch-ästhetischen Fragestellungen wieder, die ein breites, kunstinteressiertes Publikum ansprechen. Die kontinuierliche Aktivität der Kolonie fördert die Belebung der Soldiner Kiezkultur und etabliert den Wedding als Kunststandort Berlins"—or, as rendered by the English translation of the website, "The colony is understood as a space of possibilities, its projects, initiatives and collaborations are not related to marketing criteria. The heterogeneity of its members and the art shown distinguish the work of the art network. In self-organized exhibitions, discussions were with political, social and intercultural issues in artistic and aesthetic issues again that appeal to a broad audience interested in art. The continuous activity of the colony promotes the revitalization of the neighborhood Soldiner culture and established Wedding as an art venue in Berlin" (http://www.koloniewedding .de/der-verein, accessed April 25, 2015; see also http://www.koloniewedding .de/home, accessed April 25, 2015).

13. Regine Rapp and Christian de Lutz, interview with the author, May 19, 2015, Kreutzberg.

14. Tillberg had been at the Max Planck Institute in Berlin for two years. As Rapp explained to me, she and de Lutz took over "as two co-founders, two co-producers, and two co-directors, and [since] we don't have other curators, we actually are two co-curators. That means we shape and organize and do

the whole thing. So we were very clear in 2009 that we wanted to go on with interdisciplinary work, and we shaped it . . . more strongly to the now. And as not an English native I like to construct words consciously, new terms that maybe don't exist, so I call it 'nowness.' . . . We try to express and find answers for the 'nowness' by showing interesting and challenging and intelligent art pieces."

15. See https://www.cultivamoscultura.org, http://www.symbiotica.uwa.edu.au, http://bioart.sva.edu.

16. De Lutz and Rapp also aspire to create social connections that reach into their community. When I visited ALB again in the early spring of 2016, they described working on grant applications that would enable them to engage members of the growing migrant population in Berlin in their exhibitions and workshops. In the summer of 2016, at the invitation of the Alfred Erhardt Stiftung, ALB mounted *NatureCultures*, an exhibition of the work of Brandon Ballengée, Katya Gardea Browne, and Pinar Yoldas, at the Erhardt Stiftung gallery in Mitte. As this book was going to press, Art Laboratory Berlin announced that they had, for the second time, been awarded the Prize for Berlin Spaces and Initiatives, a testament to their accomplishment as a galley and workshop space.

17. For a richly illustrated guide to these exhibitions, see Rapp and de Lutz 2015.

18. These artworks were first exhibited in Houston the previous spring and then included in her volume *Greening the Galaxy* (Anker 2014).

19. These found landscapes, she points out, were by microscopic animals such as *Botryllus* and by sea sponges, ancient life forms that have possible anticancer properties of interest to contemporary biomedicine.

20. See http://aksioma.org/crustacea.deleatur, accessed June 10, 2015.

21. Rapp and de Lutz describe themselves as being interested in creating "a sustainable form of inter-disciplinarity" rather than a simple juxtaposition of science and art. Through a program of public outreach and education they hope to bridge differences in age and ability as well as expertise and discipline. As Rapp explains with reference to Smrekar's exhibit, the fact that transformation is ongoing in all of the implicated species is proof of their success as curators: "If an exhibition works both for kids, young people and special people, or a broader public, it's the proof that an exhibition is done well. If you can actually explain to a child the difference [between] a marble crayfish and the other crayfish and some we call invasive species and some we call 'glocal' or traditionally local, and then a specialist or an expert in comparative zoology might also understand it or explain it in a different way, then for us it's a proof that an exhibition has worked."

22. Brandon Ballengée, *Requilaries*, 2014, http://brandonballengee.com/projects/reliquaries, accessed May 15, 2015.

23. Pauline Beete, "Art (and Science)" Talk with Artist-Biologist Brandon Ballengée, April 6, 2015, http://arts.gov/art-works/2015/art-and-science-talk-artist

-biologist-brandon-balleng%C3%A9e. The article is also available at http://
www.feldmangallery.com/media/ballangee/general%20press/2015_ballengee
_national%20endowment%20for%20the%20arts_beete.pdf.

24. Registry of Standard Biological Parts, online database, http://parts.igem
.org/Main_Page, accessed May 18, 2015. See also the discussion of CRISPR
at the Center for Genetics and Society, http://www.geneticsandsociety.org
/index.php, and http://www.slashgear.com/crispr-cas9-modifies-your-dna
-under-legal-fire-15383737, accessed May 18, 2015; http://www.sciencealert
.com/10-things-you-need-to-know-about-the-uk-s-decision-to-allow-genetic
-modification-of-human-embryos.

25. Registry of Standard Biological Parts.

26. C-Lab (Laura Cinti and Howard Boland), "Synthetic Biology Workshop," Art
Laboratory Berlin Technical University of Berlin, September 5–7, 2014.

27. Dolphijn and Van der Tuin (2012, 7–8) affirm the "transversality of new
materialism," writing, "The strength of new materialism is precisely
this nomadic traversing of the territories of science *and* the humanities,
performing the agential or *non-innocent* nature of all matter that seems
to have escaped *both* modernist (positivist) and postmodern humanist
epistemologies."

28. Included in the show were forms from "Normal Flora," a project that considers the molds, yeasts, and bacteria that shape the scaled ecosystems that
connect our bodies to the universe; "Bed and Chair Flora"; "The Communicating Bacteria Dress," which was "created by staining textiles using pigmented
bacteria, which change colour when they send and receive communicative
signals"; works from her "Romantic Disease" series on the social, medical,
and scientific history of tuberculosis; and some works created for the Berlin show itself, including "Magic Bullet" and "The Consultation," which deal
with the role of German scientists in developing Sulfate drugs from the
dyes that they had a special expertise in creating (Rapp and de Lutz 2014,
2–3, 27).

29. The term, Haraway reminds us, was first coined by a "systems thinker,"
the landscape architect and environmentalist Beth Dempster (Haraway
2015, 260).

CONCLUSION

1. This lateral movement has been most compelling to me throughout my career, as
I explained in the introduction. Pointing to Rosi Braidotti's attention to metamorphosis and Manuel DeLanda's attention to morphogenesis, Dolphijn and
van der Tuin argue that this new materialist approach approaches "matter
[as] a transformative force *in itself*, which in its ongoing change, will not allow
any representation to take root" (Dolphijn and Van der Tuin 2012, 107). See
also Jessop and Sum. 2015.

2. "When a pattern of changes of DNA methylation is found to occur repeatedly at specific loci, discriminating the phenotypically affected cases from control individuals, this is regarded as an indication that epigenetic perturbation has taken place that is associated, possibly causally, with the phenotype. This approach is described as an epigenome-wide association study (EWAS) [7], and takes its cue from the association of genetic variability with phenotypes in genome-wide association studies (GWAS)" (Birney et al. 2016).

3. Richardson, "Plasticity and Programming: Feminism and the Epigenetic Imaginary," draft ms. courtesy of the author, 30; forthcoming in fall 2017, *Signs*. See also Richardson et al. 2014.

4. "Using a microscope where two persons could simultaneously examine the same embryo they analyzed and classified a large number of malformations caused by mutations in genes controlling early embryonic development. For more than a year the two scientists sat opposite each other examining Drosophila embryos resulting from genetic crosses of mutant Drosophila strains. They were able to identify 15 different genes which, if mutated, would cause defects in segmentation" ("The Nobel Prize in Physiology or Medicine 1995," press release, October 9, 1995, http://www.nobelprize.org/nobel_prizes /medicine/laureates/1995/press.html.

5. As Keller (1993) observed, "The simple fact is that for many years geneticists had little reason to refer to eggs and their cytoplasmic structure, and even less reason to talk about events prior to fertilization. The discourse of gene action had established a spatial map that lent the cytoplasm scientific invisibility to geneticists ('indifferent' was how Morgan described the cytoplasm) and a temporal map that defined the moment of fertilization as origin, with no meaningful time before fertilization. In this schema, there was neither time nor place in which to conceive of the egg's cytoplasm exerting its effects." See also Keller 2002b, 2004.

6. Mike Fortun (2015, 53) argues that we do well to realize that postgenomic research excites all of those affects, and encourages us to be open to finding "surprising effects driven by and embodying of attentive care." My interpretation of the complexity of Waddington's responses to the World Egg and its hatchling is in harmony with that suggestion.

7. Anne Fausto-Sterling, "How Many Sexes Are There?" *New York Times*, March 12, 1993. A29. See also Anne Fausto-Sterling 2000.

8. Azeen Ghorayshi, "Conversations with Anne Fausto-Sterling," *Method Quarterly*, no. 4 (Fall 2015), http://www.methodquarterly.com/2015/11 /conversations-with-anne-fausto-sterling.

9. As she describes the tenets of the field, "Systems thinkers consider the dynamic interactions of all the factors contributing to a particular trait of interest; these may balance one another to attain stability, or, when for some reason one or more factors change, the dynamic balancing act can destabilize a system and lead to change. Change occurs when a system first becomes

destabilized but after a time reaches some new stable state." http://www
.annefaustosterling.com/fields-of-inquiry/dynamic-systems-theory/.

10. I thank Anne Fausto-Sterling for sharing the video at https://www.youtube
.com/watch?v=PiH_S5MYHr4 with me and Sarah Richardson for telling me
about the animation.

11. http://www.annefaustosterling.com/#1446583429521-a5848f5f-070b.

12. Anne Fausto-Sterling, interview with the author, Skype, April 21, 2016.

13. Fausto-Sterling interview.

14. Fausto-Sterling interview; emphasis added.

15. "With the river thing, I was starting to write it up and I found myself using
textually the river metaphor, and then I was just stumbling over myself to
make it work, and I'm like, 'I gotta draw this.' Well, I can't draw, so I started
Googling images, and then that PDF I sent you yesterday was the result of
several days' work using Photoshop, which I'm only moderately good at, and
at one point I got it almost where I wanted it and it was a total mess, so then
I sent [it] to a person in our department who will work on figures, and she
is pretty good at Photoshop, and she improved it. Then I improved it a little
more. I'm still not sure it's where I want it. But it's good enough so that now
I think it kind of offers out the metaphor visually that I wanted" (Fausto-
Sterling interview). For another perspective on the river as metaphor that
would be fascinating to put in conversation with this use of it, see Elizabeth A.
Povinelli 2015. "Transgender Creeks and the Three Figures of Power in Late
Liberalism." *Differences* 26, no. 1: 168–87.

16. Anne Fausto-Sterling, personal communication, April 18, 2016.

17. Fausto-Sterling interview.

18. "Anastomosis," Wikipedia, https://en.wikipedia.org/wiki/Anastomosis, accessed April 21, 2016.

REFERENCES

Abouheif, Ehab, Marie-Julie Favé, Ana Sofia Ibarrarán-Viniegra, Maryna P. Lesoway, Ab Matteen Rafiqi, and Rajendhran Rajakumar. 2014. "Eco-Evo-Devo: The Time Has Come." In *Ecological Genomics: Ecology and the Evolution of Genes and Genomes*, ed. C. R. Landry and N. Aubin-Horth, 107–25. Dordrecht: Springer Netherlands.

Adams, Vincanne, Michelle Murphy and Adele E. Clarke. 2009. "Anticipation: Technoscience, life, affect, temporality." *Subjectivity* 28, 246–65.

Alaimo, Stacy, and Suzanne Heckman, eds. 2008. *Material Feminisms*. Bloomington: Indiana University Press.

Allen, Matthew. 2015. "Compelled by the Diagram: Thinking through C. H. Waddington's Epigenetic Landscape." *Contemporaneity* 4, no. 1. doi:10.5194/contemp.2015.143.

Anker, Suzanne. 2014. *The Greening of the Galaxy*. Dallas: Deborah Colton Gallery.

Appadurai, Arjun and Carol A. Breckenridge. 2009. "Foreword." In *Soak: Mumbai in an Estuary*, edited by Anuradha Mathur and Dilip da Cunha, vii-ix. New Delhi: Rupa & Co.

Arribas-Ayllon, Michael, Andrew Bartlett, and Katie Featherstone. 2010. "Complexity and Accountability: The Witches' Brew of Psychiatric Genetics." *Social Studies of Science* 40: 499–524.

Azarello, Nina. 2013. "Charles Jencks' Cells of Life Is a Manmade Landscape." *Designboom*, http://www.designboom.com/art/cells-of-life, accessed April 20, 2015.

Baedke, Jan. 2013. "The Epigenetic Landscape in the Course of Time." *Studies in History and Philosophy of Biological and Biomedical Sciences* 44: 756–73.

Barad, Karen. 2007. *Meeting the Universe Halfway: Quantum Physics and the Entanglement of Matter and Meaning*. Durham, NC: Duke University Press.

———. 2012. "Matter Feels, Converses, Suffers, Desires, Yearns and Remembers: Interview with Karen Barad." In *New Materialism: Interviews and Cartographies*, ed. Rick Dolphijn and Iris van der Tuin, 48–70. London: Open Humanities.

Birney, Ewan, George Davey Smith, and John M. Greally. 2016. "Epigenome-wide Association Studies and the Interpretation of Disease–Omics." *PLoS Genetics* (June 23). http://dx.doi.org/10.1371/journal.pgen.1006105, accessed August 10, 2016.

Bogost, Ian. 2012. *Alien Phenomenology; or, What It's Like to Be a Thing.* Minneapolis: University of Minnesota Press.

Bolling, Ruben [Ken Fisher]. 2004. "Bad Blastocyst: When Blastocysts Go Bad." Cartoon in *Thrilling Tom: The Dancing Bug Stories*, 140. New York: Andrews McMeel.

Bolter, Jay David, and Richard Grusin. 1996. "Remediation." *Configurations* 4, no. 3: 311–58.

Borges, Jorge Luis. 2015. "The Aleph" (1945). In *Grain Vapor Ray: Textures of the Anthropocene*, ed. Katrin Klingan, Ashkan Sepahvand, Christoph Rosol, and Bernd M. Scherer, 195–202. Berlin: Haus der Kulturen der Welt.

Braidotti, Rosi. 2006. *Transpositions: On Nomadic Ethics.* Cambridge: Polity.

Brockman, John. British Library. *British Comics Collection.* http://www.bl.uk, accessed February 27, 2015.

Caianiello, Silvia. 2009. "Adaptive versus Epigenetic Landscape: A Visual Chapter in the History of Evolution and Development." In *Graphing Genes, Cells, and Embryos: Cultures of Seeing 3D and Beyond*, ed. Sabine Brauckmann, Christina Brandt, Denis Thieffry, and Gerd B. Müller, 65–78. Berlin: Max-Planck-Institut für Wissenschaftsgeschichte.

Cheramie, Kristi Dykema. 2011. "The Scale of Nature: Modeling the Mississippi River." Design Observer Group, March 21. http://places.designobserver.com /entryprint.html?entry=25658, accessed September 25, 2013.

Chute, Hillary. 2010. *Graphic Women: Life Narrative and Contemporary Comics.* New York: Columbia University Press.

———. 2014. *Outside the Box: Interviews with Contemporary Cartoonists.* Chicago: University of Chicago Press.

———. 2016. *Disaster Drawn.* Cambridge, MA: Harvard University Press.

Clarke, Bruce. 2012. "From Information to Cognition: The Systems Counterculture, Heinz von Foerster's Pedagogy, and Second-Order Cybernetics." *Constructivist Foundations* 7, no. 3: 196–207.

———. 2014. *Neocybernetics and Narrative.* Minneapolis: University of Minnesota Press.

———, ed. 2015. *Earth, Life, and System: Evolution and Ecology on a Gaian Planet.* New York: Fordham University Press.

Clarke, Bruce, and Mark B. N. Hansen. 2009. "Neocybernetic Emergence." In *Emergence and Embodiment: New Essays on Second-Order Systems Theory*, ed. Bruce Clarke and Mark B. N. Hansen, 1–17. Durham, NC: Duke University Press.

Code, Lorraine. 2006. *Ecological Thinking: The Politics of Epistemic Location.* Oxford: Oxford University Press.

Cohen, Benjamin R. 2005. "Marjorie Grene." *The Believer*, March. http://www
.believermag.com/issues/200503/?read=interview_grene, accessed June 3, 2015.

Cohen, Hal. 1992. "Representation and Landscape: Drawing and Making in the
Landscape Medium." *Word and Image* 8, no. 3 (July): 243–75.

———. 2002. "Bioscience Moves into Galleries as Bioart." *The Scientist*, November 11. http://www.the-scientist.cm/?articles.view/articleNo/14362/title
/Bioscience-Moves-into-Galleries-as-Bioart.

Corner, James. 1996a. "Aqueous Agents: The (Re)presentation of Water in
the Landscape Architecture of Hargreaves Associates." *Process Architecture*
128:46–61.

———. 1996b. "Ecology and Landscape as Agents of Creativity." In *Ecological
Design and Planning*, ed. George Thompson and Frederick Steiner, 80–10. New
York: Wiley.

Crittenden, Roger. 2003. *Film and Video Editing*, 2d ed. London: Routledge.

Czerwiec, M. K., Ian Williams, Susan Merrill Squier, Michael J. Green, Kimberly R. Myers, and Scott Thompson Smith. 2015. *Graphic Medicine Manifesto*.
University Park: Pennsylvania State University Press.

Daston, Lorraine. 2010. Objectivity. Cambridge, MA: Zone.

Davis, Heather, and Etienne Turpin. 2015. *Art in the Anthropocene: Encounters
among Aesthetics, Politics, Environments and Epistemologies*. London: Open
Humanities.

DeLanda, Manuel. 2013. *Intensive Science and Virtual Philosophy*. London: Blooms-
bury Academic.

de Lauretis, Teresa. 1987. *Technologies of Gender: Essays on Theory, Film, and Fiction*. Bloomington: Indiana University Press.

Diedrich, Lisa. 2016. *Indirect Action: Schizophrenia, Epilepsy, AIDS, and the Course
of Health Activism*. Minneapolis: University of Minnesota Press.

Dolphijn, Rick, and Iris van der Tuin. 2012. *New Materialism: Interviews and Cartographies*. London: Open Humanities.

Duden, Barbara. 1993. *Disembodying Women: Perspectives on Pregnancy and the
Unborn*. Cambridge, MA: Harvard University Press.

Dumitriu, Anna. 2014. *The Bacterial Sublime [Micro]biologies Exhibition Series by
Art Laboratory Berlin*. Berlin: Blurb.

Eisenstein, Zillah R. 2001. *Manmade Breast Cancers*. Ithaca, NY: Cornell University Press.

Eisner, Will. 2004 [1985]. *Comics and Sequential Art*. Tamarac, FL: Poorhouse.

Fagan, Melinda Bonnie. 2012. "Waddington Redux: Models and Explanations in Stem Cell and Systems Biology." *Biology and Philosophy* 27: 179–213.
doi:10.1007/s10539-011-9294-y.

Fausto-Sterling, Anne. 1993. "The Five Sexes." *The Sciences* 33, no. 2 (March–
April): 20–24.

———. 2000. "The Five Sexes Revisited." *The Sciences* 40, no. 4 (July–August):
18–23.

———. 2012. "Not Your Grandma's Genetics: Some Theoretical Notes." *Psychology of Women Quarterly* 36, no. 4: 411–18.

———. 2015. "Gender as Process, not Trait: Dynamic Approaches to the Origins of Difference in Infancy." Keynote talk presented at the Foundation for Psychocultural Research Sex/Gender Conference, University of California, Los Angeles, October 23–24.

Favareau, Donald. 2010. *Essential Readings in Biosemiotics*. Springer Science and Business Media.

Feinberg, Andrew P. 2008. "Epigenetics at the Epicenter of Modern Medicine." *JAMA* 299, no. 11: 1345–50.

Fleck, Ludwik. 1986 [1929]. "On the Crisis of 'Reality.'" In *Cognition and Fact: Materials on Ludwik Fleck*, ed. Robert S. Cohen and Thomas Schnelle, 47–57. Boston Studies in the Philosophy of Science, vol. 87. New York: Springer.

———. 2012 [1935]. *Genesis and Development of a Scientific Fact*. Chicago: University of Chicago Press.

Fortun, Mike. 2015. "What Toll Pursuit: Affective Assemblages in Genomics and Postgenomics." In *Postgenomics: Perspectives on Biology after the Genome*, ed. Sarah S. Richardson and Hallam Stevens, 32–55. Durham, NC: Duke University Press.

Franklin, Sarah. 2013. *Biological Relatives: IVF, Stem Cells, and the Future of Kinship*. Durham, NC: Duke University Press.

Gilbert, Scott F. 1991. "Epigenetic Landscaping: Waddington's Use of Cell Fate Bifurcation Diagrams." *Biology and Philosophy* 6: 134–54.

———. 1994. *A Conceptual History of Modern Embryology*. Baltimore: Johns Hopkins University Press.

———. 2000. "Diachronic Biology Meets Evo-Devo: C.H. Waddington's Approach to Evolutionary Developmental Biology." *American Zoologist*, 40: 729–37.

———. 2012. "Commentary: 'The Epigenotype' by C. H. Waddington." *International Journal of Epidemiology* 41, no. 1: 20–23, doi:10.1093/ije/dyr186.

———. 2014. *Developmental Biology*, 10th ed. Sunderland, MA: Sinauer Associates.

Gilbert, Scott F., and David Epel. 2009. *Ecological Developmental Biology: Integrating Epigenetics, Medicine, and Evolution*. Sunderland, MA: Sinauer Associates.

Gilbert, Scott F., and Marion Faber. 1996. "Looking at Embryos: The Visual and Conceptual Aesthetics of Emerging Form." In *The Elusive Synthesis: Aesthetics and Science Boston Studies in the Philosophy of Science*, ed. Alfred I. Tauber, 121–51. Boston Studies in the Philosophy of Science, vol. 182. Dordrecht, The Netherlands: Kluwer Academic.

Gilbert, Scott F., Jan Sapp, and Alfred I. Tauber. 2012. "A Symbiotic View of Life: We Have Never Been Individuals." *Quarterly Review of Biology* 87, no. 4 (December): 325–41. http://www.jstor.org/stable/10.1086.668166, accessed December 19, 2012.

Gleick, James. 2011. *Chaos: Making a New Science*, revised and enhanced ed. New York: Open Road Integrated Media.

Goldberg, Aaron D., C. David Allis, and Emily Bernstein. 2007. "Epigenetics: A Landscape Takes Shape." *Cell* 128 (February 23): 635–38.

Goldberg, Michael A. 1979. "The Man-Made Future by C. H. Waddington." *Town Planning Review* 50, no. 3 (July): 369–71.

Goodall, Marcus C. 1965. *Science and the Politician*. Cambridge, MA: Schenkman Publishing Company.

Green, Louise. Forthcoming. *The Nature Industry and the Post Colony*. University Park: Penn State University Press.

Green, Michael. 2015. "Comics and Medicine: Peering into the Process of Professional Identity Formation." *Academic Medicine* 90, no. 6 (June): 774–79.

Grene, Marjorie. 1969. "Bohm's Metaphysics and Biology." *Sketching Theoretical Biology: Toward a Theoretical Biology*. Ed. C. H. Waddington, Vol. 2. *Towards a Theoretical Biology: An International Union of Biological Sciences Symposium*, vol. 2. Edinburgh: Edinburgh University Press, 61–68.

———. 1976. "Philosophy of Medicine: Prolegomena to a Philosophy of Science." *Proceedings of the Biennial Meeting of the Philosophy of Science Association*: 77–93.

———. 1995. *A Philosophical Testament*. Chicago: Open Court.

Grene, Marjorie, and David Depew. 2004. *The Philosophy of Biology: An Episodic History*. New York: Cambridge University Press.

Grusin, Richard. 2015. "Radical Mediation." *Critical Inquiry* 42 (Autumn): 124–48.

Grusin, Richard, and Jay David Bolter. 2000. *Remediation: Understanding New Media*. Cambridge, MA: MIT Press.

Halas, John, and Roger Manville. 1968. "The Animated Cartoon." *Design* 70, no. 2: 25–27.

Hall, Brian K. 1992. "Waddington's Legacy in Development and Evolution." *American Zoologist* 32:1, 113–22.

Hansson, Göran K., and Kristina Edfeldt. 2005. "Toll to Be Paid at the Gateway to the Vessel Wall." *Arteriosclerosis, Thrombosis, and Vascular Biology* 25, no. 6: 1085–87.

Haraway, Donna. 1976. *Crystals, Fabrics and Fields: Metaphors of Organicism in Twentieth-Century Developmental Biology*. New Haven, CT: Yale University Press.

Haraway, Donna, with Martha Kenney. 2015. "Anthropocene, Capitalocene, Chthulhocene." In *Art in the Anthropocene: Encounters among Aesthetics, Politics, Environments and Epistomologies*, ed. Heather Davis and Etienne Turpin, 255–70. London: Open Humanities.

Harris, Paul A. 1997. "The Itinerant Theorist: Nature and Knowledge/Ecology and Topology in Michel Serres." *SubStance* Vol. 26, no 2. Issue 83: 37–58.

Haven, Leif. 2014. "We've All Always Been Lichens: Donna Haraway, the Cthulhucene, and the Capitalocene." *Entropy*. http://entropymag.org/weve-all-always -been-lichens-donna-haraway-the-cthulhucene-and-the-capitalocene.

Henderson, Lawrence J. 1958. *The Fitness of the Environment*. Boston: Beacon Press.

Herrington, Susan. 2010. "The Nature of Ian McHarg's Science." *Landscape Journal* 29, no. 1: 1–20.

His, Wilhelm. 1885. *Anatomie menschlicher Embryonen*. vol. 3, *Zur Geschichte der Organe*. Leipzig: F. C. W. Vogel.

Holling, C. S. 1973. "Resistance and Stability of Ecological Systems." *Annual Review of Ecology and Systematics* 41: 1–23.

Hopwood, Nick. 1999. "'Giving Body' to Embryos: Modeling, Mechanism, and the Microtome in Late Nineteenth-Century Anatomy." *Isis* 90, no. 3 (September): 462–96.

———. 2000. "Producing Development: The Anatomy of Human Embryos and the Norms of Wilhelm His." Bulletin of the History of Medicine, vol. 74, no. 1, 29–79.

———. 2006. "Pictures of Evolution and Charges of Fraud: Ernst Haeckel's Embryological Illustrations." *Isis* 97, no. 2 (June): 260–301.

———. 2007. "A History of Normal Plates, Tables, and Stages in Vertebrate Embryology." *International Journal of Developmental Biology* 51, no. 1: 1–26.

———. 2015. *Haeckel's Embryos: Images, Evolution, and Fraud*. Chicago: University of Chicago Press.

Hörl, Erich. 2008. "A Thousand Ecologies: The Process of Cyberneticization and General Ecology." *Modern Language Notes* 123: 194–217.

Hunemann, Philippe, and Maël Lemoine. 2014. "Introduction: The Plurality of Modeling." *History and Philosophy of the Life Sciences* 36, no. 1: 5–15. doi:10.1007/s40656-014-0002-5.

Iberall, A. S. "A Personal Overview." Waddington, ed. 1969. *Towards a Theoretical Biology: An International Union of Biological Sciences Symposium*, vol. 2. Edinburgh: Edinburgh University Press. 10–17.

Import Projects. 2014. Press Release, Multitudes. http://import-projects.org/images/pressreleases/import_projects_multitudes_en.pdf.

Jablonka, Eva, and Marion J. Lamb. 2002. "The Changing Concept of Epigenetics." *Annals of the New York Academy of Sciences* 981: 82–96.

Jablonka, Eva, Marion J. Lamb, illustrated by Anna Zeligowski. 2014. *Evolution in Four Dimensions: Genetic, Epigenetic, Behavioral, and Symbolic Variation in the History of Life*. Cambridge, MA: MIT Press.

Jacob, François, David Perrin, Carmen Sanches, and Jacques Monod. 2005 [1960]. "The Operon: A Group of Genes Whose Expression Is Coordinated by an Operator." *Comptes rendus hebdomadaire des séances de l'Academie des sciences* 250, no. 6 (facsimile): 1727–29.

Jamniczky, Heather A., Julia C. Boughner, Campbell Rolian, Paula N. Gonzalez, Christopher D. Powell, Eric J. Schmidt, Trish E. Parsons, Fred L. Bookstein, and Benedikt Hallgrìmsson. 2010. "Rediscovering Waddington in the Post-Genomic Age." *Bioessays* 32: 553–58. doi:10.1002/bies.200900289.

Jantsch, Erich, and C. H. Waddington. 1976. *Evolution and Consciousness: Human Systems in Transition*. Reading, MA: Addison-Wesley.

Jencks, Charles. 2003. *The Garden of Cosmic Speculation*. London: Frances Lincoln.

———. 2011. *The Universe in the Landscape: Landforms by Charles Jencks*. London: Frances Lincoln.

Jessop, Bob, and Ngai-Ling Sum. 2015. "Pre-disciplinary and Post-disciplinary Perspectives." *New Political Economy* 6, no. 1: 89–10. doi:10.1080/13563460020027777.

Jones, P. Lloyd. 1971. "Waddington, C. H. (1969) Behind Appearance." *British Journal for the Philosophy of Science* 22, no. 2 (May): 183–87.

Keller, Evelyn Fox. 1993. "Rethinking the Meaning of Genetic Determinism." Tanner Lecture on Human Values delivered at the University of Utah, February 18. http://tannerlectures.utah.edu/_documents/a-to-z/k/keller94.pdf.

———. 1996. "Drosophila Embryos as Transitional Objects: The Work of Donald Poulson and Christiane Nüsslein-Volhard." *Historical Studies in the Physical and Biological Sciences* 26, no. 2: 313–46.

———. 2000. "Models of and Models for: Theory and Practice in Contemporary Biology." *Philosophy of Science* 67, no. 3: S72–S86.

———. 2003 [2002] *Making Sense of Life: Explaining Biological Development with Models, Metaphors, and Machines*. Cambridge, MA: Harvard University Press.

———. 2004. "What Impact, if Any, Has Feminism Had on Science?" *Journal of Biosciences* 29, no. 1 (March): 7–13.

———. 2010. *The Mirage of a Space between Nature and Nurture*. Durham, NC: Duke University Press.

———. 2014. "From Gene Action to Reactive Genomes." *Journal of Physiology* 592, no. 11: 2423–29.

———. 2015. "The Postgenomic Genome." In *Postgenomics: Perspectives on Biology after the Genome*, ed. Sarah S. Richardson and Hallam Stevens, 9–31. Durham, NC: Duke University Press.

Kelty, Chris, and Hannah Landecker. 2004. "A Theory of Animation: Cells, L-systems, and Film." *Grey Room*, no. 17 (Fall): 30–63.

Kemp, Martin. 1996. "Doing What Comes Naturally: Morphogenesis and the Limits of the Genetic Code." *Art Journal* 55, no. 1 (Spring): 27–32.

Klopmeier, Kurt. 2015. " 'Wherein Past, Present, and Future He Beholds': Comics and the Eternal Present." *Critical Flame*, issue 36 (May–June). http://criticalflame.org/comics-and-the-eternal-present, accessed June 7, 2015.

Knight, Paula. 2013. *X-Utero: A Cluster of Comics*. Bristol, UK. Unpaginated.

———. 2017 [2016]. *The Facts of Life*. University Park: Penn State University Press.

Komara, Ann E. 2000–2001. "The Glass Wall: Gendering the American Society of Landscape Architects." *Studies in the Decorative Arts* 8, no. 1 (Fall–Winter): 22–30.

Kramer, Peter D. 2014. "Why Doctors Need Stories." *New York Times*, October 19, SR1.

Landecker, Hannah. 2011. "Food as Exposure: Nutritional Epigenetics and the New Metabolism." *BioSocieties* 6, no. 2: 167–94.

———. 2013. "Metabolism, Reproduction, and the Aftermath of Categories." *Scholar and Feminist Online* 11, no. 3 (Summer). http://sfonline.barnard.edu /life-un-ltd-feminism-bioscience-race, accessed August 10, 2016.

Latour, Bruno. 1987. "The Enlightenment without the Critique: A Word on Michel Serres' Philosophy." *Royal Institute of Philosophy Lectures Series* 21: 83–97. doi:10.1017/S0957042x00003497.

———. 1996. *Aramis, or the Love of Technology*, Cambridge: Harvard University Press.

———. 2004. "Why Has Critique Run out of Steam? From Matters of Fact to Matters of Concern." *Critical Inquiry* 30 (Winter): 225–48.

———. 2014. "Agency at the Time of the Anthropocene." *New Literary History* 45, no. 1 (Winter): 1–18. doi:10.1353//nlh.2014.0003.

Leka, Kaisa. 2008. *I Am Not These Feet*. Helsinki: Absolute Truth.

Leopold, Luna B. 1994. *A View of the River*. Cambridge, MA: Harvard University Press.

Levy, Roy, and Ron Rappaport. 1982. "Gregory Bateson 1904–1980." *American Anthropologist* 84, no. 2 (June). http://www.interculturalstudies.org/Bateson /biography.html, accessed August 10, 2016.

Lie, Merete, and Nina Lykke. 2016. *Assisted Reproduction across Borders: Feminist Perspectives on Normalizations, Disruptions, and Transmissions*. London: Routledge.

Lippit, Yukio. 2012. "Of Modes and Manners in Japanese Ink Painting: Sesshū's *Splashed Ink Landscape* of 1495." *Art Bulletin* 94, no. 1: 50–77.

Locke, Margaret. 2013. "The Lure of the Epigenome." *The Lancet* 381, no. 9881 (June): 1896–97.

Lohier, Patrick. 2014. "*Here* Is Richard McGuire's Epic of Time." Boingboing.net, December 11. http://boingboing.net/2014/12/11/here-is-richard-mcguires-epi .html, accessed June 8, 2015.

Loi, Michele, Lorenzo Del Savio, and Elia Stupka. 2013. "Social Epigenetics and Equality of Opportunity." *Public Health Ethics* 6, no. 2: 142–53.

Love, Alan C. 2010. "Idealization in Evolutionary Developmental Investigation: A Tension between Phenotypic Plasticity and Normal Stages." *Philosophical Transactions of the Royal Society B* 365: 679–90.

MacDonald, Amanda. 1996. "Bandes Dessinées, or Patches of Culture." *Australian Journal of French Studies* 33, no. 2 (May): 185–203.

———. 1998. "In Extremis: Hergé's Graphic Exteriority of Character." *Other Voices* 1, no. 2 (September). http://www.othervoices.org/1.2/amacdonald/herge .php.

Maienschein, Jane. 2014. *Embryos under the Microscope: The Diverging Meanings of Life*. Cambridge, MA: Harvard University Press.

Malabou, Catherine. 2006. "One Life Only: Biological Resistance, Political Resistance," trans. Carolyn Stead. *Critical Inquiry* 42 (Spring): 429–38.

Mandala Collaborative. 1975. "Pardisan: Plan for an Environmental Park in Tehran." http://www.annewhistonspirn.com/pdf/Pardisan_Plan.pdf, accessed September 4, 2015.

———. 1976. "The Pahlavi Environmental Prize." *Environmental Conservation* 3, no. 1: 58. doi:10.1017/s0376892900017781.

Margulis, Lynn. 1995. "Gaia Is a Tough Bitch." In *The Third Culture: Beyond the Scientific Revolution*, ed. John Brockman, 129–46. New York: Simon and Schuster.

Margulis, Lynn, James Corner, and Brian Hawthorne, eds. 2007. *Ian McHarg: Conversations with Students/Dwelling in Nature*. New York: Princeton Architectural Press.

Margulis, Lynn, and Dorion Sagan. 2002. *Acquiring Genomes: A Theory on the Origin of Species*. New York: Basic.

Mathur, Anuradha, and Dilip da Cunha. 2001. *Mississippi Floods: Designing a Shifting Landscape*. New Haven, CT: Yale University Press.

———. 2006. *Deccan Traverses: The Making of Bangalore's Terrain*. New Delhi: Rupa.

———. 2009. *Soak: Mumbai in an Estuary*. New Delhi: Rupa.

———. 2014a. "Waters Everywhere." In *Design in the Terrain of Water*, eds. Anuradha Mathur, Dilip da Cunha, et al. 1–13. Philadelphia: Applied Research & Design Publishing.

———. "Interview with the author," March 5, 2015. Skype.

Mathur, Anuradha, Dilip da Cunha, Caitlin Squier-Roper, Graham Laird Prentice, and Matthew J. Wiener. 2014b. "Structures of Coastal Resilience." Structures of Coastal Resilience website, July 17. http://structuresofcoastalresilience.org, accessed June 19, 2015.

———, eds., 2014b. *Design in the Terrain of Water*, Philadelphia: Applied Research & Design Publishing.

———. 2014c. "Turning the Frontier: Norfolk and Hampton Roads, Virginia." *Structures of Coastal Resilience Phase I: Context, Site, and Vulnerability Analysis*. Structures of Coastal Resilience." Structures of Coastal Resilience website, July 17. http://structuresofcoastalresilience.org, accessed February 26, 2017.

———. 2014d. "Norfolk VA Strategies and Design." *Structures of Coastal Resilience Phase 2: Detailed Design Executive Summary*. Structures of Coastal Resilience website, 17 June 2014. http://structuresofcoastalresilience.org/wp-content/uploads/2014/06/SCR_Phase_2_Summary.pdf.

McCloud, Scott. 1994. *Understanding Comics: The Invisible Art*. New York: William Morrow Paperbacks.

McGuire, Richard. 2014. *Here*. New York: Pantheon.

McHarg, Ian. 1969. *Design with Nature*. Garden City, NY: Natural History Press.

———. 2006. *Design with Nature*. Wiley Series in Sustainable Design. New York: John Wiley & Sons, Inc. [Kindle edition].

———. 2007a. "Pond Water." In *Ian McHarg: Conversations with Students/Dwelling in Nature*, ed. Lynn Margulis, James Corner, and Brian Hawthorne, 77. New York: Princeton Architectural Press.

———. 2007b. "The Theory of Creative Fitting." In *Ian McHarg: Conversations with Students/Dwelling in Nature*, ed. Lynn Margulis, James Corner, and Brian Hawthorne, 20–61. New York: Princeton Architectural Press.

Meloni, Maurizio, and Giuseppe Testa. 2014. "Scrutinizing the Epigenetics Revolution." *BioSocieties* 9, no. 4: 431–56. doi:10.1057/biosoc.2014.22.

Meyer, Elizabeth. 1997. "The Expanded Field of Landscape Architecture." In *Ecological Design and Planning*, ed. George F. Thompson and Frederick R. Steiner, 45–79. New York: John Wiley.

———. 2011. "The Expanded Field of Landscape Architecture (Excerpt)." *Landscape Urbanism* 1 (Summer). http://scenariojournal.com/landscape-urbanism-journal-archive.

Mitchell, Robert. 2010. *Bioart and the Vitality of Media: The Cultural Mediations of Bioscience*. Seattle: University of Washington Press.

Morgan, Lynn M. 2004. "A Social Biography of Carnegie Embryo No. 836." *Anatomical Record* 276B: 3–7.

———. 2009. *Icons of Life: A Cultural History of Human Embryos*. Berkeley: University of California Press.

Mukherjee, Siddhartha. 2016a. "The Improvisational Oncologist." *New York Times Sunday Magazine*, May 15, 43.

———. 2016b. "Same but Different: Identical Twins and the Science of Epigenetics." *New Yorker*, May 2: 24–30.

Mullins, Phil. 2000. "Vintage Marjorie Grene." *Tradition and Discovery* 27 (1): 33–45.

———. 2010. "Marjorie Grene and Personal Knowledge." *Tradition and Discovery* 37, no. 2:20–44.

Murphy, Michele. 2013. "Distributed Reproduction, Chemical Violence, and Latency." *Scholar and Feminist Online* 11, no. 3 (Summer). http://sfonline.barnard.edu/life-un-ltd-feminism-bioscience-race/distributed-reproduction-chemical-violence-and-latency, accessed August 5, 2016.

Myers, Natasha. 2015. *Rendering Life Molecular: Models, Modelers, and Excitable Matter*. Durham, NC: Duke University Press.

Nerlich, Brigitte. 2015. "Tools for Thinking about an Increasingly Complex World." http://blogs.nottingham.ac.uk/makingsciencepubic/files/2013/12, accessed April 10, 2015.

Niemann, Christopher. 2015. "A Rose." *New Yorker*, March 16. http://www
.newyorker.com/culture/culture-desk/cover-story-2015–03–23, accessed August 11, 2016.

Nilsen, Anders. 2014. "Me and the Universe." *New York Times*, September 24.
http://www.nytimes.com/interactive/2014/09/25/opinion/private-lives-me
-and-the-universe.html.

Nilsson, Lennart. 1990. *A Child Is Born*. New York: Delacourt Press.

Nixon, Rob. 2006–2007. "Slow Violence, Gender, and the Environmentalism of
the Poor." *Journal of Commonwealth and Postcolonial Studies* 13, no. 2–14, no. 1:
14–37.

Noble, Denis. 2011. "A Theory of Biological Relativity: No Privileged Level
of Causation." *Interface Focus* 2, no. 1 (February 6): 55–64. doi:10.1098/
rsfs.2011.0067.

Noomin, Diane. 2012. "Baby Talk: A Tale of 4 Miscarriages." In *Glitz-2-Go*. Seattle:
Fantagraphics.

Nugent et al. 2015. "Brain Feminization Requires Active Repression of Masculin-
ization via DNA Methylation." *Nature Neuroscience* 18(5): 690–97.

Oestreicher, Christian. 2007. "Basic Research: A History of Chaos Theory." *Dia-
logues in Clinical Neuroscience* 9, no. 3: 279–89.

Oyama, Susan. 1985. *The Ontogeny of Information: Developmental Systems and
Evolution*. Cambridge: Cambridge University Press.

———. 2000. *Evolution's Eye: A Systems View of the Biology-Culture Divide*. Durham, NC: Duke University Press.

———. 2015. "Sustainable Development: Living with Systems." In *Earth, Life, and
System: Evolution and Ecology on a Gaian Planet*, ed. Bruce Clarke, 203–24. New
York: Fordham University Press.

Oyama, Susan, Paul E. Griffiths, and Russell D. Gray. 2001. "Introduction: What
is Developmental Systems Theory?" In *Cycles of Contingency: Developmental
Systems and Evolution*, ed. Oyama, Griffiths and Gray, 1–11. Cambridge, MA:
MIT Press.

Panovsky, Aaron. 2015. "From Behavior Genetics to Postgenomics." In *Postgenomics: Perspectives on Biology after the Genome*, ed. Sarah S. Richardson and Hallam
Stevens, 150–73. Durham, NC: Duke University Press.

Pardee, Arthur B. 2001. "Ruth Sager 1918–1997." In *Biographical Memoirs*, vol. 80,
2–15. Washington, DC: National Academy Press.

Park, Robert Ezra. 1936. "Human Ecology." *American Journal of Sociology* 42, no. 1
(July): 1–15.

Parkes, Graham. 2011. "Japanese Aesthetics." *Stanford Encyclopedia of Philosophy*,
ed. Edward N. Zalta. http://plato.stanford.edu/archives/win2011/entries
/japanese-aesthetics, accessed August 11, 2016.

Parnes, Ohad. 2007. "Die Topographie der Vererbung: Epigenetische Land-
schaften bei Waddington und Piper." *Trajekte* 14, no. 7 (April): 26–32.

―――. 2015. "The Epigenetic Landscape" translation by Ohad Parnes, manuscript courtesy of the author.

Paul, David L., Amitai Abramovitch, Scott L. Rauch, and Daniel A. Geller. 2014. "Obsessive-Compulsive Disorder: An Integrative Genetic and Neurobiological Perspective." *Nature Reviews Neuroscience* 15 (June): 410–22.

Peck, A. L., trans. 1949. *Aristotle's De Generatione Animaleum.* Cambridge, MA: Harvard University Press. http://archive.org/stream/generationofanimooarisuoft /generationofanimooarisuoft_djvu.txt.

Pevzner, Nicholas, and Sanjukta Sen. 2010. "Preparing Ground: An Interview with Anuradha Mathur and Dilip Da Cunha." *Places: Design Observer*, June 29. http://places.designobserver.com/feature/preparing-ground-an-interview -with-anuradha-mathur-and-dilip-da-cunha/13858, accessed November 4, 2012.

Pigliucci, Massimo. 2013. "On the Different Ways of 'Doing Theory' in Biology." *Biological Theory*, July 19: Volume 7, Issue 4, 287–97. doi:10.1007s13752–012–0047–1.

Piper, John. 1937. "Lost, A Valuable Object." In *The Painter's Object*, ed. Myfanwy Piper, 69–73. London: Gerald Howe.

Povinelli, Elizabeth A. 2015. "Transgender Creeks and the Three Figures of Power in Late Liberalism." *Differences* 26, no. 1: 168–87.

Rapp, Regine, and Christian de Lutz. 2014. "Introduction." In *Anna Dumitriu: The Bacterial Sublime*, 2–4. Berlin: Blurb.

―――. 2015. Interview with the author, May 19, 2015, Kreutzberg, Berlin.

―――, eds. 2015. *[Macro]biologies and [Micro]biologies: Art and the Biological Sublime in the 21st Century.* Berlin: Blurb.

Reed, Chris, and Nina-Marie Lister. 2014. "Ecology and Design: Parallel Genealogies." *Places Journal*, April. https://placesjournal.org/article/ecology-and -design-parallel-genealogies, accessed April 15, 2015.

Rheinberger, Hans-Jörg. 2010. *An Epistemology of the Concrete: Twentieth-Century Histories of Life.* Durham, NC: Duke University Press.

Richardson, Sarah S. 2015. "Maternal Bodies in the Postgenomic Order: Gender and the Explanatory Landscape of Epigenetics." In *Postgenomics: Perspectives on Biology after the Genome*, ed. Sarah S. Richardson and Hallam Stevens, 210–31. Durham, NC: Duke University Press.

―――. 2017. "Plasticity and Programming: Feminism and the Epigenetic Imaginary." Forthcoming, *Signs.*

Richardson, Sarah S., Cynthia R. Daniels, Matthew W. Gillman, Janet Golden, Rebecca Kukla, Christopher Kuzawa, and Janet Rich-Edwards. 2014. "Don't Blame the Mothers." *Nature* 512 (August 14): 131–32.

Richardson, Sarah S., and Hallam Stevens, eds. 2015. *Postgenomics: Perspectives on Biology after the Genome.* Durham, NC: Duke University Press.

Robertson, Alan. 1977. "Conrad Hal Waddington. 8 November 1905–26 September 1975." *Biographical Memoirs of Fellows of the Royal Society* 23 (November): 575–622.

Ruse, Michael. 1990. "Are Pictures Really Necessary? The Case of Sewell Wright's 'Adaptive Landscapes.'" *Proceedings of the Biennial Meeting of the Philosophy of Science Association* 2: 63–77.

Rybczynski, Zbig. 1997. Rybczynski, Zbigniev. "Looking to the Future: Imagining the Truth." *Cinema & Architecture: Melies, Mallet-Stevens, Multimedia* (1997): 182–98.

Sagan, Dorion. 2007. "'The Algae Will Laugh': McHarg and the Second Law of Thermodynamics." In *Ian McHarg: Conversations with Students/Living in Nature*, ed. Lynn Margulis, James Corner, and Brian Hawthorne, 79–94. New York: Princeton Architectural Press.

———, ed. 2012. *Lynn Margulis: The Life and Legacy of a Scientific Rebel*. White River Junction, VT: Chelsea Green.

Sante, Luc. 2015. "Richard McGuire's *Here*." *New York Times*. October 15.

Schaechter, Moselio. 2012. "Lynn Margulis (1938–2011)." *Science* 335, no. 6066 (January 20): 302. doi:10.1126/science.1218027.

Schmitt, Gail K. 2008. "Sager, Ruth." In *Dictionary of Scientific Biography*, vol. 24, 316–21. New York: Charles Scribner's Sons.

Scott, Joan Wallach. 2004. "Feminism's History." *Journal of Women's History* 16, no. 2, 10–29.

Serres, Michel, and Bruno Latour. 1995. *Conversations on Science, Culture, and Time*. Ann Arbor: University of Michigan Press.

Serres, Michel, Josué V. Harari, and David F. Bell. 1982. *Hermes—Literature, Science, Philosophy*. Baltimore: Johns Hopkins University Press.

Serres, Michel, Sheila Faria Glaser, and William R. Paulson. 1997. *The Troubadour of Knowledge*. Ann Arbor: University of Michigan Press.

Shapiro, James. 2015. "Bringing Cell Action into Evolution." In *Earth, Life and System: Evolution and Ecology on a Gaian Planet*, ed. Bruce Clarke, 175–202. New York: Fordham University Press.

Shortland, Michael. 1998. "Michel Serres, Passe-partout." *British Society for the History of Science* 31, no. 3 (September): 335–53.

Slack, Jonathan M. W. 2002. "Conrad Hal Waddington: The Last Renaissance Biologist?" *Nature Reviews/Genetics* 3 (Nov.): 889–95.

Small, David. 2009. *Stitches: A Memoir*. New York: W. W. Norton.

Smrekar, Maja. 2014. "Aquatic Art Laboratories: Native and Invasive Species." Gallery talk. Art Laboratory Berlin. July 20, 2014. Transcription by the author.

———. 2015. "Aquatic Art Laboratories—on Native and Invasive Species." In *[Macro]biologies and [Micro]biologies: Art and the Biological Sublime in the 21st Century*, ed. Regine Rapp and Christian de Lutz, 41–45. Berlin: Blurb.

Sousanis, Nick. 2015. *Unflattening*. Cambridge, MA: Harvard University Press.

Spalding, Frances. 2009. *John Piper, Myfanwy Piper: Lives in Art*. Oxford: Oxford University Press.

Spirn, Anne Whiston. 1976. "Pardisan Park." *Environmental Conservation* 3, no. 1: 58. doi:10.1017/s0376892900017781.

———. 2000. "Ian McHarg, Landscape Architecture, and Environmentalism: Ideas and Methods in Context." In Michel Conan, *Environmentalism in Landscape Architecture*, 97–114. Washington, DC: Dumbarton Oaks Research Library and Collection.

Squier, Susan Merrill. 2004. *Liminal Lives: Imagining the Human at the Frontiers of Biomedicine*. Durham, NC: Duke University Press.

———. 2008. "'So Long as They Grow Out of It': Comics, The Discourse of Developmental Normalcy, and Disability." *Journal of the Medical Humanities* 29: 71–88.

———. 2011. *Poultry Science, Chicken Culture: A Partial Alphabet*. New Brunswick, NJ: Rutgers University Press.

———. 2013. "The World Egg and the Ouroboros." *Scholar and Feminist Online* 11, no. 3. http://sfonline.barnard.edu/life-un-ltd-feminism-bioscience-race, accessed July 23, 2014.

———. 2015. "The Uses of Graphic Medicine for Engaged Scholarship." In *Graphic Medicine Manifesto*, by M. K. Czerwiec, Ian Williams, Susan Merrill Squier, Michael J. Green, Kimberly R. Myers, and Scott T. Smith, 41–66. University Park: Pennsylvania State University Press.

Stevens, Hallam and Sarah S. Richardson. 2015. "Beyond the Genome." In *Postgenomics: Perspectives on Biology after the Genome*. Durham: Duke University Press, 1–8.

Stewart, Kathleen. 2008. "Weak Theory in an Unfinished World." *Journal of Folklore Research* 45, no. 1 (January–April): 71–82.

Stiegler, Bernard. 2012. "Lights and Shadows in the Digital Age." Keynote lecture, Digital Inquiry Symposium. Berkeley Center for New Media. http://bcnm.berkeley.edu/2012/05/02/digital-inquiry-keynote-talk-by-bernard-steigler/, accessed February 25, 2017.

Stiegler, Bernard and Irit Rogoff. 2010. "Transindividuation." *E-flux* 14 (March): 1–6. http://www.e-flux.com/journal/14/61314/transindividuation/, accessed September 17, 2012.

Streeter, George L. 1942. "Developmental Horizons in Human Embryos: Description of Age Group XI, 13 to 20 Somites and Age Group XII, 21 to 29 Somites." *Contributions to Embryology* 30: 211–45.

Strick, James. 2015. "Exobiology at NASA: Incubator for the Gaia and Serial Endosymbiosis Theories." In *Earth, Life, and System: Evolution and Ecology on a Gaian Planet*, ed. Bruce Clarke, 80–104. New York: Fordham University Press.

Sullivan, Shannon. 2013. "Inheriting Racist Disparities in Health: Epigenetics and the Transgenerational Effects of White Racism." *Critical Philosophy of Race* 1, no. 2: 190–218.

Tavory, I., E. Jablonka and S. Ginsburg. 2012. "Culture and Epigenesis: A Waddingtonian View." In *Oxford Handbook of Culture and Psychology*, ed. J. Valsiner, 662–676. New York: Oxford University Press.

Tavory, Iddo, L. S. Ginsburg, and Eva Jablonka. 2013. "The Reproduction of the Social: A Developmental System Approach." In *Developing Scaffolds in Evolution, Culture, and Cognition*, ed. L. Caporael, J. Griesemer, and W. Wimsatt, 307–25. Cambridge, MA: MIT Press.

Thelen, Esther. 1995. "Motor Development: A New Synthesis." *American Psychologist* 50: 79–95.

Thom, René. 1969. "Topological Models in Biology." *Topology* 8: 313–35.

———. 1970. "Topological Models in Biology." In Waddington, ed. *Towards a Theoretical Biology: An International Union of Biological Sciences Symposium*, vol. 3, 89–116. Edinburgh: Edinburgh University Press.

———. 1980. "Halte au hazard, silence au bruit." *Le Debat*. 3:3, 119–32.

———. "From Animal to Man: Thought and Language." In Favareau, Donald. 2010. *Essential Readings in Biosemiotics*, 337–76. New York: Springer Science and Business Media.

Thom, René and Robert E. Chumley. 1983. "Stop Chance! Silence Noise!" *SubStance* 12, no. 3, issue 40, "Determinism," 11–21.

Towler, Mhairi, and Paul Harrison. 2012. "C. H. Waddington: Inspiring New Creations," November 23. In *Towards Dolly: Edinburgh, Roslin and the Birth of Modern Genetics*. http://libraryblogs.is.ed.ac.uk/towardsdolly, accessed August 11, 2016.

Tuana, Nancy. 2008. "Viscous Porosity: Witnessing Katrina." In *Material Feminisms*, ed. Stacy Alaimo and Susan Hekman, 188–213. Bloomington: Indiana University Press.

US Army Corps of Engineers. 2013. Coastal Risk Reduction and Resilience, CWTS 2013-3, Directorate of Civil Works, US Army Corps of Engineers, Washington DC.

Vaccari, Andrés, and Belinda Barnet. 2009. "Prolegomena to a Future Robot History: Stiegler, Epiphylogenesis and Technical Evolution." *Transformations: Journal of Media, Culture, and Society*, no. 17. http://www.transformationsjournal.org/journal/issue_17/article_09.shtml, accessed April 19, 2015.

Van Speybroeck, Linda. 2002. "From Epigenesis to Epigenetics: The Case of C. H. Waddington." *Annals of the New York Academy of Sciences* 981: 61–81.

Van Speybroeck, Linda, Dani De Waele, and Gertrudis Van De Vijver. 2002. "Theories in Early Embryology: Close Connections between Epigenesis, Preformationism, and Self-Organization." *Annals of the New York Academy of Sciences* 981: 7–49.

Vasiliou, Stella K. and Eleftherios P. Diamandis, moderators, with George M. Church, Henry T. Greely, Francoise Baylis, Charis Thompson, and Gerold Schmitt-Ulms. 2016. *Clinical Chemistry*. 62–10: 1304–11.

Vogt, Günter. 2008. "The Marbled Crayfish: A New Model Organism for Research on Development, Epigenetics and Evolutionary Biology." *Journal of Zoology* 276: 1–13. doi:10.1111/j.1469-7998.2008.00473.x.

Waddington, C. H. 1940. *Organisers and Genes*. Cambridge: Cambridge University Press.

———. 1942. "Canalisation of Development and the Inheritance of Acquired Characters." *Nature* 150, no. 3811: 563–65. doi:10.1038/150563a0.

———. 1948 [1941]. "Art between the Wars." In *The Scientific Attitude*, ed. C. H. Waddington, 36–52. West Drayton, Middlesex, UK: Penguin.

———. 1948. *The Scientific Attitude*. West Drayton, Middlesex, UK: Penguin.

———. 1957. *The Strategy of the Genes: A Discussion of Some Aspects of Theoretical Biology*. London: Allen and Unwin.

———. 1961 [1951]. "The Character of Biological Form." In *Aspects of Form: A Symposium on Form in Nature and Art*, ed. Lancelot Law Whyte, 43–52. Bloomington: Indiana University Press.

———. 1968a. "Towards a Theoretical Biology." *Nature* 218 (May 11): 525–27.

———, ed. 1968b. *Towards a Theoretical Biology: An I.U.B.S. Symposium*, vol. 1. Edinburgh: Edinburgh University Press.

———, ed. 1969. *Towards a Theoretical Biology: An I.U.B.S. Symposium*, vol. 2. Edinburgh: Edinburgh University Press.

———, ed. 1970. *Towards a Theoretical Biology: An I.U.B.S. Symposium*, vol. 3. Edinburgh: Edinburgh University Press, 1970.

———. 1970 [1969]. *Behind Appearance: A Study of the Relations between Painting and the Natural Sciences in This Century*. Cambridge, MA: MIT Press.

———, ed. 1972. *Towards a Theoretical Biology: An I.U.B.S. Symposium*, vol. 4. Edinburgh: Edinburgh University Press.

———. 1973. "Operational research in World War II: OR against the U-boat." London: Elek Books.

———. 1977. *Tools for Thought: How to Understand and Apply the Latest Scientific Techniques of Problem Solving*. New York: Basic.

———. 1978. *The Man-Made Future*. New York: Springer.

———. 2010. "The Practical Consequences of Metaphysical Beliefs on a Biologist's Work: An Autobiographical Note." In Waddington, ed., *Sketching Theoretical Biology: Toward a Theoretical Biology*, vol. 2, 79–81. New Brunswick, NJ: Transaction.

———. 2012. "The Epigenotype." International Journal of Epidemiology 41 (1): 10–13. [1942] *Endeavor* 1: 18–20.

———. 2014. *The Strategy of the Genes*. Routledge Library Editions: 20th Century Science. [Kindle edition].

Waldheim, Charles. 2002. "How Do You Draw a River? Review of *Mississippi Floods: Designing a Shifting Landscape*." *Landscape Journal* 20, no. 2: 220–23.

Weideman, Paul. "Negotiating Man and Nature." *The Santa Fe New Mexican*, May 5, 2006, Pasatiempo; Pg. PA-24 sec.

Wellmann, Janina. 2011. "Introduction: Science and Cinema." *Science in Context.* 24, no. 3 (September 2011): 311–28.

———. 2017. *The Form of Becoming: Embryology and the Epistemology of Rhythm, 1760–1830.* Boston: Zone.

Williams, Ian. 2014. *The Bad Doctor.* University Park: Pennsylvania State University Press.

Woolf, Virginia. 2007. *To the Lighthouse: Selected Works of Virginia Woolf.* Herfordshire, UK: Wordsworth.

Woolston, Chris. 2016. "Researcher under Fire for *New Yorker* Epigenetics Article." *Nature* 583, no. 295 (May 9). doi:10.1038/533295f.

Zimmer, Carl. 2016. "Growing Pains for the Field of Epigenetics as Some Call for Overhaul." *New York Times*, July 1.

Design in the Terrain of Water (Mathur and da Cunha), 168, 233n6

Design with Nature (McHarg), 138–39, 142, 146, 150

determinism, 1–2, 6–8, 15, 18, 38, 46, 59–60, 208

"Developmental Organization Purpose and Ethics" (Waddington), 228n1

developmental systems theory, 31, 185, 221n5

Dickinson, Emily, 150

Diedrich, Lisa, 218n21

Disney, Walt(er), 87

Dolphijn, Rick, 14, 237n1

drawing. *See* art; comics; epistemology; graphic medicine

Dreyfus, Hubert, 39, 43, 220n14

Duany, Andreas, 233n5

Dumitriu, Anna, 201–3, 214

Ecological Design and Planning (Meyer), 165–66

ecology, 18, 42–43, 129–30, 133–55, 161–73, 193–95

Edwards, Robert, 65

Einstein, Albert, 54

Eisner, Will, 95–97

Eliasson, Olafur, 136

Eliot, T. S., 27

Elsasser, W. M., 39

embodiment, 44, 77–78, 92–93, 116–18, 146, 173, 198–99, 229n3

embryology, 15, 21–22, 35–36, 65–66, 69–84, 87–99, 101, 131–32, 157, 175, 205–10, 218n19, 225n1, 228n11, 232n3. *See also* world egg

EMBRYO project, 225n1

entanglement, 18, 37–38, 109, 151, 186, 195, 207

EpiGeneSys, 126–27

epigenetics: art and, 183–203; ball on the hill model of, 11, 28–29, 69–71; controversies over, 9–10; creativity and, 15–16; definitions of, 1–2,

6, 37–38, 184, 215n2, 216n9, 217n15, 218n19, 219n5; determinism and, 1–2; embryology and, 15, 17, 21–22, 35–36, 65–66, 69–84, 87–99, 207–10; ethics and, 186–87; gender and, 8, 10, 16, 18–19, 30–32, 201–2, 205–6, 210–11; graphic medicine and, 99–121; heritability and, 9–10; landscape architecture and, 129–37, 232n3; metaphysics of, 37–39; modeling and, 2, 18, 27–28, 28, 45–49, 60–67, 69–84; pregnancy concerns and, 116–21; river model of, 11, 21–24, 161–68; scaling and, 11–12, 54, 60–65, 135–37, 205, 207; temporality and, 11–12, 105, 111–13, 206–7; theoretical biology and, 16, 30, 34–39, 41–49, 51–52, 57–61, 129, 217n13; transdisciplinarity and, 3, 7–13, 61–65, 69–71, 84–85, 121–30, 142–50, 168–73, 210–12; view from underneath with guy wires model of, 11, 15, 18, 184, *185*, 187, 202–3

epistemology: comics and, 99–121, 125–26; design and, 129–37, 152–55; drawing (as practice) and, 144, 152–55; embodiment and, 77–78, 92–93, 108–9, 116–18, 146, 173, 198–99, 229n3; entanglement and, 18, 37–38, 109, 151, 186, 195, 207; feminist, 89–90, 116–18, 134, 165–68, 173, 186–87, 190–91; McHarg and, 139–55; morality and, 109–10; reductionism and, 18, 22, 30–43, 111, 147, 156–57; rhythm and, 84–85, 95, 224n25, 229n3. *See also* metaphysics; ontologies; process philosophy

ethics, 14, 91, 99–102, 186–87, 194, 198, 228n1

evidence-based medicine, 91

evo-devo-eco, 8, 114

evolution, 8, 28, 30, 35, 41, 47, 53, 61–62, 76–77, 83, 96, 105, 112–13, 143–49, 155, 195, 203, 219n7, 227n21, 229n2, 230n17, 235n10